PRAISE FOR
THE
SONG OF
OUR
SCARS

"In this insightful and humane book about pain, suffering, and survival, Haider Warraich once again braids history and personal history to confront questions both ancient and contemporary. It is a marvelous read."
—Dr. Siddhartha Mukherjee, author of *The Emperor of All Maladies*

"Pain is both a universal experience and one that is deeply connected to class, gender, race, and power—truths that the opioid epidemic and declining life expectancy have made tragically apparent. In *The Song of Our Scars*, Warraich explores how the medical community's approach to pain went off the rails and makes a passionate case for more holistic, person-centered treatment. Beautifully written and deeply humane, this is essential reading for anyone seeking to understand the roots of the opioid crisis." —Beth Macy, author of *Dopesick*

"As physician-author Warraich aptly points out, 'Almost everything we know about pain and how we treat it is wrong.' His masterful new book is a unique, panoramic, and deep view of pain, taking us through his personal experience, its history and evolution, the science, and the massive corporate corruption that caused the opioid epidemic. An incredible book." —Dr. Eric Topol, author of *Deep Medicine*

"A wonderful exploration of the chronic pain conundrum—past and present—in all of its varied dimensions: biomedical, psychological, social, and economic. Warraich is truly a gifted storyteller."

—Dr. Damon Tweedy, author of *Black Man in a White Coat*

"*The Song of Our Scars* is a brilliant deep dive into the emotional, physical, and metaphysical world of pain. The very personal and ultimately hopeful book takes us on a remarkable journey across millennia and deep into the brain and consciousness. It also offers an unsparing look at how the cure (for chronic pain) became the disease. Dr. Haider Warraich has written *The Emperor of All Maladies* for pain."

—Dr. Jonathan Reiner, author of *Heart*

"Pain is the most universal yet misunderstood aspect of what it means to be human. Dr. Warraich leverages his own intimate relationship with suffering and highly developed skills as a physician healer to pen a masterful book that will change your lens on mankind forever. *The Song of Our Scars* attacks the principal element of the human condition to be deciphered if we hope to mitigate the torment of societal scourges such as racism, poverty, chronic illness, and loneliness. The net result of Dr. Warraich's research and writing will, I believe, shorten the distance from our brains to our hearts and create a path toward healing and lasting empathy for one another."

—Dr. Wes Ely, author of *Every Deep-Drawn Breath*

"To paraphrase Virginia Woolf, considering how common pain is, it's strange that it hasn't 'taken its place with love and battle and jealousy among the prime themes of literature.' *The Song of Our Scars* goes a long way to remedying this. Essential reading for anyone hoping to understand what it means to be alive."

—Daniel Wallace, author of *Big Fish*

THE
SONG OF
OUR
SCARS

THE UNTOLD STORY OF PAIN

HAIDER WARRAICH

BASIC BOOKS

New York

Basic Books
Hachette Book Group
1290 Avenue of the Americas, New York, NY 10104
www.basicbooks.com

Printed in the United States of America

First Edition: April 2022

Published by Basic Books, an imprint of Perseus Books, LLC, a subsidiary of Hachette Book Group, Inc. The Basic Books name and logo is a trademark of the Hachette Book Group.

The Hachette Speakers Bureau provides a wide range of authors for speaking events. To find out more, go to www.hachettespeakersbureau.com or call (866) 376-6591.

The publisher is not responsible for websites (or their content) that are not owned by the publisher.

Print book interior design by Marie Mundaca

Library of Congress Cataloging-in-Publication Data
Names: Warraich, Haider, author.
Title: The song of our scars: the untold story of pain / Haider Warraich.
Description: First edition. | New York: Basic Books, 2022. | Includes bibliographical
    references and index.
Identifiers: LCCN 2021038446 | ISBN 9781541675308 (hardcover) | ISBN 9781541675292 (ebook)
Subjects: LCSH: Warraich, Haider—Health. | Pain—History. | Pain—Treatment. | Pain—Social
    aspects. | Chronic pain—Physiological aspects. | Chronic pain—Psychological aspects. |
    Chronic pain—Patients—United States—Biography.
Classification: LCC RB127 .W37 2022 | DDC 616/.0472—dc23/eng/20211029
LC record available at https://lccn.loc.gov/2021038446

ISBNs: 9781541675308 (hardcover), 9781541675292 (ebook)

LSC-C

Printing 1, 2022

For my favorite girls
Ammi, Belu, Evoo, and Gogo

CONTENTS

INTRODUCTION

There is no coming to consciousness
 without pain.

— Carl Jung

PAIN IS A FUNDAMENTAL TRUTH. Pain might well be the first sensation a baby feels as it's born, a gateway to the world of conscious experience, almost certainly becoming the sensation it most strongly associates with being alive. And indeed, every subsequent day of our lives, we experience pains of different types. These are often innocuous but can at times become intractable. Pain is one of the most consistent aspects of the consensus reality we all experience, a hallmark of consciousness among all beings, hardwired into our frames through evolutionary mechanisms millions of years in the making.

Yet pain is also the most fluid of all sensations. While how we see, hear, touch, and taste has likely remained unaffected by historical changes, how we perceive and tolerate pain has changed considerably just in the last century. Pain has transformed from a spiritual force, often the only language through which celestial agents could speak to mortal beings, into a corporeal corruption that can be entirely comprehended and conquered with biomedical advances. Yet other aspects of the place of pain in our society have remained unchanged. Pain—in how it is recognized, treated, and inflicted—has always been and remains an instrument of power, often used against the weak. For it is impossible to separate the assessment of pain from the assignment of supremacy.

Pain is imperialistic: European colonists often derided the pain of their Black and Brown subjects, chalking it up to feebleness, even as they capitalized on its affliction. As Britain operated the greatest opioid-production machine in the history of mankind, waging war simply to keep selling opium and addicting foreign populations to it, it banned the use of opiates for its own people, knowing just how addictive the poppy can be.

Pain is racial: Black slaves were often subjected to indescribable violence under the false pretext that they were too numb to feel pain the way their white masters did. Even to this day, otherwise sophis-ticated people, including some physicians, hold on to antiquated fabrications, including that Black people feel less pain because their skin is thicker than white people's. This is one reason why the pain of Black people remains both underrecognized and undertreated.[1]

Pain is gendered: women are more likely to feel pain, but their pain is also more likely to be dismissed. Many women who seek relief are belittled and delegitimized by some of the very doctors they turn to for support.

And most of all, pain is personal, so personal that it is said to be the one thing truly our own, so inscrutable that it cannot even be communicated within the constraints of language. The only reason I mustered the gumption to write this book, to attempt to uncover the nature of our most complicated sensation, is that pain has been a part of my being for almost my entire adult life.

Pain is as sure a sign of life as the beating of the heart; its disappearance often signals death. It touches us from the tips of our toes to the crowns of our heads; it can even be felt in long-departed parts of our body, as in the phantom pains of those who have lost their limbs to landmines, diabetes, or flesh-eating bacteria. It is tempting to describe the evolution of weapons, from rocks and pointy branches to armed drones and tactical nuclear missiles, in terms of their ability to kill.

But their primary aim is often to inflict pain—physical, psychological, cultural, racial, and economic.

And yet there are also those for whom pain is a ladder to the divine. In many religions, such as the Shiite sect of Islam, self-flagellation is a core ritual because of its supposedly redemptive qualities. Connoisseurs of spicy food set their palates on fire as they reach for peppers with ever-rising Scoville units, relishing the bodily chaos that ensues. For some, pain provides a sexual thrill that no amount of gentle stimulation can achieve.

Pain is also our most effective teacher. Most lessons fade, most sermons go unheard, but those taught by pain can last a lifetime. My baby daughter might not listen to anything I say, but the hot skillet only has to instruct her once for her never to touch it again. Not every lesson pain teaches is worth remembering though. Corporal punishment might be a poor tool for education, but it is very effective at creating lifelong trauma.

In the last two centuries, our understanding of how our bodies flourish and falter has advanced tremendously. And yet, even as the song of our scars reaches a deafening pitch, pain remains the sensation we comprehend the least.

It is not an accident that we fail to understand the very basics of pain—especially its more entrenched manifestation, chronic pain. The attempt to define pain beginning in the nineteenth century using clinical and scientific terms shrank its scope to fit the constraints of the tools and rituals of medicine. The corporatization of healthcare transformed people into consumers, transmuting human suffering into a lucrative opportunity to maximize capital. And profit-hungry pharmaceutical companies hijacked a movement to provide comfort to people with terminal illnesses, resulting in one of the most carefully engineered campaigns of distortion in human history.

As a physician I treat people who hurt every day. Yet my relationship with pain goes back to before I was the one people turned to for respite.

When I was a medical student in Pakistan, I would spend hours every day grinding at the gym. If I wasn't playing basketball, I was lifting weights or running on the treadmill. I was no star athlete, but I labored joyously like a hamster on its wheel. Exercise was medicine for my body, therapy for my spirit.

One evening in the middle of a bench press, I heard a loud click in my back. All at once, my body went limp, and the metal bar I was holding aloft came crashing down, pinning me to the bench. Panic set in. With my torso compressed under more than two hundred pounds of steel, I began to suffocate. I tipped the bar over to one side, hoping the weights would slide off, but I quickly remembered that I had them clamped in place.

Only then did I do what I should have done many long milliseconds before. With progressively thinning breaths, I cried for help.

A couple of medical students rushed over and lifted the weights awkwardly off my chest. Like me, they came to the air-conditioned sports center not just to escape the blistering Karachi sun but to zone out from the incessant memorization of pathologies and the drugs used to treat them. And yet my injury had turned their sanctuary into one of the bays in the emergency room close by. They grabbed me by my arms and tried to pull me up to my feet, but I screamed out in agony. One of them rushed outside and brought back a wheelchair.

Prior to that evening, I had made the walk from the hospital to the gymnasium several hundred times. It was short and typically forgettable. But I will always remember that ride back to the medical center in the wheelchair. Every small bump, even the fine clefts between pavement panels, shuddered through my body.

This was the day that pain became a part of my life. It changed my line of work. It took many of my friends from me. It also took away those precious hours at the gym, my sanctuary from the demands of medical school. And it could have taken away even more: more than once, my pain led me to believe that the only way to release myself from its vise would be to end my life.

On that day in the gym, pain settled in as a persistent presence in my body, an infestation that would come to shape the narrative arc of my life and a primary reason I wrote this book. And yet I know that my story is in fact quite unremarkable. Nothing is as important to our ability to survive as our ability to hurt.

Because I was a medical student who knew most of the staff, when I reached the emergency room, I was immediately taken to a secluded bed without having to linger in the overcrowded waiting room. The pain was most severe in my back, but it traveled all the way up and down my spine like a pendulum, shattering everything in its path. The emergency room physician jabbed me with a dose of intravenous ketorolac, an anti-inflammatory painkiller, and told me that I had sprained a muscle. It would get better in a week, he reassured.

That week went by, and then another, and then several more. My discomfort completely incapacitated me. Sitting hurt too much, but standing wasn't any better. I could only sleep by lying on my side in the fetal position, with my knees up to my chest, a pillow in between them.

Walking made me exquisitely aware of how my body shifted its weight between the legs and the lower back. There wasn't a special maneuver I could perform to be comfortable. There was no escape hatch. When I had previously torn my bicep muscle while lifting weights, it hurt only when I moved my arm in certain ways. Those movements I could easily avoid. Hurting my back offered no such out.

Sitting, standing, and walking: as a physician in training, these three activities constituted almost every waking moment of my day. When I was supposed to be attending to my patients during rounds with my team, I could think only of trying to find a chair to sit in or a wall to lean against. Of all the places in the hospital, I dreaded being in the operating room the most, as it meant standing possibly for hours. Medical students are usually assigned a position far from the action, performing tasks whose only purpose can seem to be to torture. I would find myself contorting at awkward angles, holding metal retractors to keep gaping incisions in the body open. I was supposed to be watching the surgeon's hands, but all I could see was the side of the table or one of the operators' backs. It quickly became clear that surgery was not a viable career path for me. Being a doctor required me to be empathic; yet my ceaseless pain had left me looking only inward, as I constantly introspected every aberrant alarm arising within my body. In my darkest days, I wasn't even sure I could ever practice medicine at all.

Chronic pain has governed my life ever since. To this day, an atypical movement or a long day at work threatens to put me back at the beginning of this treacherous trail. But my story is not special, my adversity not exceptional. The back is in fact the part of our body we most commonly injure. And while most back injuries tend to heal, many become indomitable, making them the most common cause of both disability and chronic pain in the United States and around the world. And given that the condition is relatively invisible, with no bleeding wounds or bulbous tumors, and that most of us don't really think about our backs until they bother us, only those who live with back pain can truly grasp what it wreaks.[2]

Many people's lives will, at some point, be completely reorganized by chronic pain. Chronic pain is a truly global phenomenon, estimated to affect 1.5 billion people. Even so, it is a malady with a distinctly American bent: one study shows that Americans appear to feel aches and pains more frequently than people in other countries, and Americans are most likely to use opioids for their pain. One in five American adults—an estimated sixty-six million in total—experience chronic pain, resulting in $500 billion in direct medical costs and lost productivity. Chronic pain disproportionately affects those who are already disadvantaged in other ways: women, people of color, the poor, the elderly, the unemployed, and those living in rural areas are all more likely to experience it. Almost twenty-four million Americans are unable to participate in major life activities due to chronic pain. And the number of people who live with chronic pain is rising.[3]

We have always hurt. But how we hurt today—how it affects our lives, how we give it meaning, and how we attempt to overpower it—is nothing like how we have suffered in the past. Medical science is just one reason why how we hurt has changed: the confluence of several broader social, cultural, and economic movements has fundamentally altered how those of us alive today experience pain. We must learn to reckon and reconcile with the body on fire.

Whether due to an aching knee or a throbbing head, pain has long been viewed as the work of supernatural forces strumming inflamed nerves like sitar strings. Pain always had a meaning, a greater context. According to the Bible, the pain of childbirth occurred because of Eve's succumbing to the charms of the devil, which meant that until at least the eighteenth century in Western Europe, attempts to relieve labor pangs were punishable by death. The crime committed by the very first witch burned in Scotland was an attempt to ease the passage of a woman's twins; the mother, too, was set alight at the stake. Pain relief was seen as an unnatural interruption of cosmic commandments.

As Western societies secularized, so did pain. The treatment of pain moved from the chapel to the clinic, and we began to seek prescriptions rather than prayers to ease the suffering. Yet, even as anesthesia and morphine were discovered in the nineteenth century, the primary aim of medicine continued to be prolonging life rather than providing relief.

In this, medicine succeeded. Yet the longer we survived, the more years we lived with disability and in pain. As noncommunicable diseases such as cancer and heart failure replaced infectious diseases as the leading causes of death, many people's lives ended in excruciating agony, their bones moth-eaten by malignant tumors, their lungs flooded with their own secretions. Even this gruesome sight would not move physicians to ease the plight of their patients.

The twentieth century saw the rise of a broad movement that sought to give people more control over their bodies. This movement was manifested in many forms: it was cast in the women's rights movement, the civil rights movement, and decolonization. And while medicine, given its long history of paternalism, was slower to change, these broader forces began to shift power from doctors toward patients, giving people more agency over their health. This shift, first manifesting within healthcare as the hospice movement founded by Cicely Saunders in the United Kingdom, implored physicians to avert not just death but despair and discomfort. Pain was no longer viewed as a symptom of sickness but understood as a syndrome in itself— not a metaphysical disturbance caused by the tipping of the balance between virtue and sin but a purely physical sensation caused by an anatomic disturbance that could be fixed as easily as a broken bone could be put back together. Yet pain is far more complex than that simplistic view, and this turn in the cultural understanding of pain did nothing to abate the rising scourge of chronic pain in the United States and around the world.

Perhaps the most visible sign of this transformation is the opioid epidemic. The Centers for Disease Control and Prevention estimates that between 1999 and 2019, this misery was triggered by a flood of opioid prescriptions to patients with chronic pain, many of whom went on to become addicted and some of whom began abusing street drugs like heroin and fentanyl. The public reaction to the opioid epidemic has focused mostly on the pharmaceutical companies that pushed these pills. Particularly intense attention has been laid on members of the Sackler family, the owners of Purdue Pharma, which manufactured OxyContin, the prescription opioid widely believed to have triggered the crisis.

Little attention, however, is paid to the entire healthcare system that abetted Purdue Pharma and other opioid manufacturers: the physicians who sold their sacrosanct status in exchange for profits, the medical journals that published highly flawed research, the federal regulators who failed to protect the public from dangerous drugs, the distributors that flooded every bit of the country with them, and the pharmacies that asked few questions before dispensing large quantities of dangerous pills to vulnerable individuals. An entire field of medicine centered on pain was created with scant evidence guiding its practice. The corporatization of medical care, coupled with the emergence of consumerism, created a pill-popping culture that placed all of our hopes and dreams for relief on drugs and procedures.

Yet, as we recognize the broad extent of this extraordinary tragedy, we have to consider something even more elemental, something at the heart of what we need to accomplish if we are to beat back the current opioid epidemic and prevent another from recurring in the future: almost everything we know about pain and how we treat it is wrong.

Pain is sensitive to context in a way no other human sensation is. The aches that engulf a marathoner's entire body change meaning instantly when that marathoner crosses the finish line. During sex, pain can induce euphoria or leave lifelong scars. The pain in the belly that the devout believer feels during a religious fast is very different from that felt by someone starving because they cannot afford a meal. While some variation in pain sensitivity between individuals occurs due to biological differences, our response to pain is largely mediated by the surrounding circumstances we feel it in. In fact, even the genes that cause us to feel pain are deeply affected by the environments we grow up in.

We have been trying to understand the complexities of pain as long as we have been around. In Book 9 of *The Republic*, Socrates, speaking to his brother Glaucon, asks, "Do we not say that pain is opposite to pleasure?" To Socrates, pain was not merely the absence of pleasure; it was based on where a person started from: Did they move from pleasure to pain or simply from the tranquil middle? An individual's baseline sets up the subsequent experience of either pain or pleasure, which Socrates considered to be on the opposite ends of a spectrum. Moving from pleasure to pain, according to this dictum, is more distressing than moving from pain to *more* pain. Are we any closer to understanding pain today?

The expression people in pain most dread hearing is that their agony is "all in their head." It is often used to diminish that agony, to erase their very personhood. Yet our brain does have a central role in shaping how we hurt. After a pain signal reaches the brain, it undergoes significant reprocessing. The brain, based on previous experiences and current expectations, can modulate pain to be felt either more or less severely. How much something hurts can vary depending on factors like one's mood and level of distraction. The human brain is not just staffing the ticketing booth at the circus— it is the ringleader. Without the brain's permission, no tigers jump through burning hoops, no trapeze artists fly around, no swords are swallowed.[4]

The brain is particularly involved in the experience of chronic pain. Most people assume that when pain lasts long enough, usually more than three months, it transforms into chronic pain. This is how people like me who live with pain have come to think of it and how doctors also like me have been trained to treat it: chronic pain is essentially acute pain prolonged. Yet, if you dive deep into the science only now emerging, a very different picture comes together: acute and chronic pain are entirely distinct phenomena, and there is no justification for treating them the same way.

The siloed nature of science means that at present there is no single working theory for what chronic pain is. Yet my journey poring over thousands of research studies and speaking to dozens of experts and patients, as well as my own odyssey aboard this faltering body, makes one thing clear: most chronic pain is not just a physical sensation. To our nervous system, chronic pain is most often akin to an emotion we feel in a part of our body, an overlearned traumatic memory that keeps ricocheting around in our brains, often long after the injury it rehearses has fully healed. Unlike acute pain, which ascends up the spinal cord to the brain from a nick on the shin or a frayed nerve in the foot, chronic pain descends down from the brain, often with no need for an incitement from below.

Pain remains a difficult matter to approach primarily because of the gulf between the people who experience it, the clinicians who treat it, and the researchers who study it. This gulf means that many common notions are about as grounded in reality as magical fairy dust. It is said that what doesn't kill you makes you stronger. Yet research actually suggests the opposite, that people who have chronic pain are even more sensitive to pain than those for whom pain is a stop but not a terminal destination. It is said that pain is inevitable but suffering is optional. Yet I cannot imagine that anyone who has experienced persistent pain personally or cared for someone in unflagging agony would ever believe such hogwash. When pain arrives and refuses to leave, suffering is as inevitable as death itself. The only thing I have gained from pain is the lived experience so essential to knowing how ceaseless suffering can wring a human spirit dry.

Standing on this island, I feel pain coming at me from all directions. It comes to me from within when I wake up in the middle of the night in affliction. It comes to me from my patients, who often have disease eating their insides out, leaving them up to their nostrils in pain. And it has come to me from the papers, books, and articles I have read and the scientists, clinicians, and patients I have spoken to about all that hurts.

To understand pain is to know the human body and the human mind and how they are interweaved. It is the strongest riposte to how clinical medicine artificially divides them.

To understand pain is to recognize how race, gender, ethnicity, and power come to indelibly mark what it means to inhabit the human frame.

To understand pain is to learn how the greatest medical tragedy in history came to be, how corporate greed and academic naivete and corruption fueled the opioid epidemic, and how it could recur again.

To understand pain is to explore the true nature of human suffering, how religion and spirituality have often been our most potent balms, and how movements such as existentialism, feminism, and consumerism have changed not only our core beliefs but also our senses.

Yet the need to understand pain is not just a scientific curiosity for me. It has given deeper meaning to the pain that I feel. And while this knowledge may not always grant me complete relief, it has allowed me to find a new way to live in my body.

Broadening the lens through which I see pain has helped to defog the window through which I see the pain of others, an essential part of my work as a clinician. Even as physicians and nurses are almost constantly face-to-face with suffering, that existential leap known as empathy, the act of feeling another's pain, is especially important as we increasingly rely on blood tests and imaging to tell us what ails a patient, while pain continues to elude such quantification. Relieving pain is one of the most gratifying feelings a clinician can ever get in the course of their vocation. Yet the drive to label pain as a physical sensation, as a vital sign similar to heart rate or blood pressure, was not an organic movement founded by clinicians. It was in fact engineered by the pharmaceutical industry, appealing to clinicians' deep desire to relieve suffering to sell trillions of dollars' worth of products. And while the drugs and devices they sold provided comfort in the short term, for people with chronic pain they were ineffective for most and deadly for many.

If you twist your ankle or bump your head, or if you live with torment that never ebbs, what you feel and how you respond is not just the aggregate of nerve signals bombarding your brain stem. It is the sum product of your entire existence and the entire history of human beings encapsulated in the multidimensional experience we call pain.

Reaching a new understanding of how we hurt will change how we live with our aching selves. Synthesizing our knowledge about the fundamentals of pain could move us closer to a future in which even if we hurt, we don't suffer. And recognizing the many layers of pain and how we respond to the agony of others could lay the foundation for a just and equitable society.

1.

THE INTERPRETATION OF AGONY

What We Talk About When We Talk About Pain

Suffering is the entrance to the
    person. It is the door to some-
    thing much larger.
                              —Eric Cassell

THE WAY MOST PEOPLE SPEAK about pain is very different from the way doctors are trained to speak about it. One might presume that doctors would be better at talking about pain, given how central it is to their lifework, or at least more accurate. But I don't think this is necessarily true. Pain is complex, with physical and mental dimensions that both overlap and diverge. In some cases, it can be meaningless and transitory, like from the bruise you get when a heavy book falls on your toe. At other times, it can open the door to lifelong suffering, like when a tumor begins to eat into your bones. This multidimensionality of pain seems intuitive. Yet medicine has largely lost the ability to think in these terms.

According to Google, use of the word "pain" has doubled in English-language books since the 1970s, but what it has come to mean has changed. Increasingly, medicine, and by extension society as a whole, understands pain as a strictly physical sensation. This essentializing of pain to a purely mechanical disruption—stripped of its emotional, spiritual, contextual, and traumatic layers—has allowed it to become wholly medicalized. And doctors, with their outsize influence on human bodies, are partly responsible for that.

In my first few years of medical school, I was taught how pain is sensed and processed within our body. I learned about it sitting in lecture halls or poring over textbooks in the library. Later, as I began to interact with actual patients, almost everything I learned related to the assessment of people's discomfort. The prototypical patient I was trained to take a medical history from or perform a physical exam on was a person in pain, shaping the way I walk my hand across someone's belly or simply have a conversation with them to this day. Pain is the frame through which all medical students all over the world learn to see the world as physicians.

Yet, despite its centrality, how clinicians conceptualize pain and actualize that in practice is deeply flawed. Having boxed pain in as a physical entity, medicine has turned it into a vital sign, a number similar to body temperature or heart rate. This categorization we know now is not only wrong but dangerous: a study published in the journal *Anesthesia & Analgesia* in 2005 showed that assessing pain on a numeric scale doubles the risk of opioid overdoses in the hospital.

In reality, pain is one of three distinct but overlapping phenomena. On one end of the spectrum is something called nociception. Nociception, the process through which potentially injurious stimulation is detected and transmitted by our nervous system, is often confused with pain. Nociception is a purely physical sensation, shared by all living things, even cells and plants. On the other end is what many come to call suffering, an entirely psychological consequence of the experience of visceral injury or emotional trauma. Suffering is an entirely psychological experience that appears to be felt only by advanced animals, humans in particular. Between nociception and suffering is pain, both sensation and emotion, both physiological and psychological, which is most often caused by nociception and frequently causes us to suffer.

A journey that ends with a shared understanding of what pain is must begin with ensuring we are all speaking the same language. We will start by exploring nociception, the subconscious noxious signals traversing our body. For all pain and suffering stem from the same source: touch.

Anyone in agony believes that what they feel is distinctly their own, incommunicable and inscrutable. Yet the ability to feel discomfort is shared across the spectrum of life. And the study of other living beings doesn't just provide insights into what they experience; it highlights that the capacity to hurt is not exclusive to humanity and that we have much in common with the other creatures we share the earth with.

The chief need of any organism is to process information from its environment. The means by which we detect and synthesize this information—sight, taste, smell—are called sensations. And the most basic of these sensations is touch: for even a simple single-celled organism to survive, it has to know if it is in contact with something else. Yet touch is also our least understood sense, and we are only just beginning to feel it out. In fact, the two scientists who made seminal discoveries revealing how we sense touch, heat, and nociception only received the ultimate recognition in science—the Nobel Prize—in October 2021.

Touch allows us not only to find our place in the world we inhabit but to regulate the world within ourselves. Biting into an apple, feeling it fill our stomach, and then knowing when it stretches our rectum and we have to go to the restroom are all processes mediated by the ability of our cells to detect pressure and tension.

We can do this thanks to Piezo proteins, which were discovered in 2010 by Ardem Patapoutian, one of the two scientists who won the Nobel Prize in Physiology or Medicine in 2021. Named for the Greek word for pressure, these massive proteins sit across the membranes of cells. They are shaped like propellers, with three blades and a pore in the middle that can open in response to a mechanical force. When open, the pore allows ions to enter the cell, altering its charge and sending an electrical signal shooting up the nerves and spinal cord all the way to the brain, where the signal is processed.[1]

Even plants are exquisitely sensitive to touch. While certain exotic florae, such as the Venus flytrap, might be more obviously responsive to it, almost every plant has the ability to detect mechanical contact. One experiment found that plants reacted significantly to scientists merely stroking them from base to tip just once a week. Some plant species bloomed better than before and were free of pests, while others were ravaged to extinction. Another experiment on the thale cress—a small but resilient flowering plant that often grows on road-sides, up walls, and between rocks—showed that just thirty minutes of touching can change 10 percent of the plant's genome and launch a cascade of plant hormones. This might be a form of self-defense: it would allow plants to respond if insects landed on them or if other plants were growing too close to them and intruding on their share of sunlight.[2]

Given that plants exhibit extensive electrical and chemical signaling and respond to external elements like light, water, and, of course, touch, some scientists claim that plants may in fact demonstrate "intelligence." The obvious next question is whether plants that detect and react to touch can also experience what that touch feels like. What does a rose feel when you pluck its petals? What does a blade of grass perceive when it is chomped by a grazing cow?[3]

Plants are much more aware of and responsive to their environments than most of us would imagine, and their reaction to touch may feel eerily similar to our own. An insect crossing the threshold of a Venus flytrap will cause it to snap its floral jaws shut. But when anesthetic agents that put humans to sleep, such as ether, are applied to the mouth of a Venus flytrap, an insect can walk across it without eliciting any reaction. Some plants feel pressure using the same mechanically activated channels that are also present in humans. And when plants are stressed, they release the gaseous hormone ethylene, which, among other things, was used as an early-twentieth-century anesthetic agent. So when plants feel pressure, they respond in adaptive ways to protect themselves in the moment as well as in the long term. Roots respond to the soil as they grow, while winds cause stems to grow thicker and stronger.[4]

We know that ripping a fruit off a branch causes a biological reaction, but does it cause an emotional one? Probably not. A dismembered plant feels nothing beyond the inert act of mechanical force. This is because plants do not have a nervous system the way animals do, which might provide them the ability to generate feelings and consciousness. So even as touch elicits an intricate physiological response in plants, it is highly unlikely that it sparks a sentimental one.

For more complex organisms, touch spans a spectrum: a warm embrace can easily turn into a bear hug. What makes a caress of the cheek different from a smack, turning touch into nociception, is more than the intensity of the stimulus. Our unique sensitivity to touch, our ability to imbue it with meaning, represents the origin of pain.

Every sensation we feel spans a spectrum. The music of King Crimson may be transcendent to one, torture to another; the spice of habanero life affirming to one, caustic to another. The difference between the searing heat of the desert sun and the warm embrace of a fireplace lies not just in their intensity. Heat and warmth differ in their character, in their valence.

And that's the difference between a sensation and a perception: perception is the brain's way of providing context to what we sense. That's what happens when touch turns unpleasant: it goes from sensation to perception as touch transforms into nociception, the sensory nervous system's response to stimuli that damage or threaten to damage the organism.

Nociception is one of life's most ancient endowments. The ability to detect threats in one's environment is central to evading danger. It is as essential for a cell avoiding an acidic medium as it is for a person sidestepping away from a speeding motorcycle. And because simple organisms like bacteria lack the intricate set of sensations and abilities to communicate that more complex organisms have, they rely even

THE SONG OF OUR SCARS

more on physical contact to discern the world around them, similar to how a blind person might use a white cane to feel out their world.

For humans, simply the velocity of a touch can determine how pleasant it is felt to be. Humans perceive being caressed at three centimeters per second to be generally more pleasant than touch at velocities both faster and slower than that. While many researchers have hypothesized that we have wholly different systems for detecting touch that provides comfort versus discomfort, recent work by researchers at the University of Liverpool in the United Kingdom casts doubt on such a dichotomy. There is wide variation among people who rate the same graze as pleasant or unpleasant. In addition to touch, the skin has mechanisms to detect various other forms of stimuli—heat, cold, and chemical irritants—all of which can swing from a positive to a negative extreme.

For nociception to transform into pain, the organism needs to possess the ability to feel something, recognize it as noxious, assign a negative connotation to it, physically withdraw to avoid it, and learn to avoid it in the future. This transformation happens not in the skin but in our brains. Pain is in essense the conscious manifestation of nociception.

Defining consciousness remains one of the unconquered peaks in science and philosophy. Consciousness is something anyone reading these words possesses yet no one fully grasps. But, while we may not completely understand consciousness, we can still define it operationally: consciousness is quite simply an entity's unique and subjective awareness of itself and its environment. And what could be more unique and subjective than the ability to feel pain, learn from it, and express it through one's actions? Pain therefore can be considered a hallmark of consciousness. Which organisms possess consciousness and how they come to possess it could inform us about who can feel pain and how it's felt.

Two theories on the origin of consciousness have more adherents than others: the global neuronal workspace and integrated information theories. Both have holes—the integrated information theory, for example, would grant consciousness to any object, even your phone, provided it can process and integrate information—and if history provides any clue, neither will have the final word on the substance of consciousness.

Almost everyone who studies consciousness, however, can agree on the following:

1. Our experience of living, the feeling of being alive and everything that comes with it—urges and restraints, sights and sounds, memories and dreams, hindsight and foresight, pain and pleasure—is all centered in the back of the cerebral cortexes that form the gnarly crown of the human brain.

2. Not everything we sense reaches our consciousness. In fact, most of our sensations go on to live and die in our subconscious. Most of the time this is a passive process: as you text on your phone while driving, your ears don't stop hearing the other cars around you. Your eyes still catch blurry glimmers of things hurtling by, but your conscious awareness only has space for your thumb operating the phone. The collection, transmission, and delivery of neurologic signals can occur entirely outside our conscious awareness.

3. The parts of the brain that receive sensory information—such as, say, the primary visual cortex that processes what you see—don't actually contribute to the formation of the conscious experience of seeing. The reception of color and light from the eyes is not the same as their eventual representation in our minds. When you look around you, what you see is not so much the world as your re-creation of it. Our senses compress an infinitely complex universe into a simplified package that we can wrap our heads around.

4. What imbues disparate bits of electric currents encoding colors, sounds, smells, and pressure with emotions and memories is integration. Integration leads to consciousness being a singular, fluid whole.

It is through this framework of consciousness that we can understand pain and how it is different from nociception.

Nociception is the noxious signal emitted as soon as the tiny needle of the flu shot pierces the skin of your shoulder and sheds its chemical load into the surrounding muscle. Nociception is the buzz of electricity that bombards our brain when that punch lands in our gut.

Pain represents the integration of those signals in the cerebral cortexes. Pain is the interpretation of nociception, translating its wordless language into a vocabulary that we can understand. Many factors affect this transformation. Two are intensity and velocity: the gentle touch of a feather can easily turn noxious if the same feather is mashed into your skin. Another is preference: spicy food triggers the same receptors as tear gas; yet those who enjoy spice perceive those two sensations in very different ways. Integration with other senses also plays a part: knowing that the hand caressing your back is that of a loved one will arouse comfort, while that of a stranger might cause revulsion.

The coming to consciousness of pain turns subconscious nociception into a conflagration, as the spark traveling down the cord detonates the dynamite. In fact, the link between pain and consciousness is so integral in part because the same segments of the brain generate both processes.[5]

The central problem of pain in recent times is that it has been conflated with nociception; people have come to believe that pain is synonymous with the sensory signals coming to your brain from distant parts of the body that are hurting. But in fact pain is a conscious, integrated experience that the human mind generates, most often (but not always) in response to nociceptive overload. Feeling pain is not like feeling other sensory elements. Heat and cold receptors in the skin are specially designed for one purpose, directly responding to hot or cold temperatures and communicating their presence or absence to the brain, just as taste receptors directly respond to sugar hitting the tongue. But pain is not an element that exists around us.

Pain was first defined by the International Association for the Study of Pain (IASP) in 1979 as "an unpleasant sensory and emotional experience associated with actual or potential tissue damage, or described in terms of such damage." While this definition came to be accepted widely, there was controversy surrounding the final clause. It is not at all clear that one needs to be able to describe pain to experience it. Even the slightest sliver of consciousness is enough for an organism to experience pain. Since the definition was published, we have learned that pain can be felt even by those in a coma, although they may not be able to describe or remember it. "We haven't kept up with advances in pain," said Srinivasa Raja, an anesthesiologist and pain researcher at Johns Hopkins, who goes by Raj. "Some of the criticism was that the definition didn't adequately address pain in disempowered or neglected populations such as newborns and the elderly."[6]

The IASP asked Raj to lead a committee that would be responsible for coming up with a new definition of pain. "I did not anticipate the importance of the task and how globally the prior definition had been accepted and how much thought had gone into it," he told me. "Now I think of it as a legacy of sorts." After years of deliberations, including input from the public, the committee came up with a version very similar to the previous one but without the requirement to be able to describe it. They defined pain as "an unpleasant sensory and emotional experience associated with, or resembling that associated with, actual or potential tissue damage." Removing the need to be able to describe pain has extended it from something that only exists in our fully conscious state to something associated with any level of consciousness, even when, say, we are in a coma and our existence is fully limited to the subconscious.[7]

As the difficulty in coming up with a definition for it shows, pain is an almost indescribably individual feeling that eludes any quantification. It is informed in its every variance by personal histories, cultural idiosyncrasies, racial and gender imbalances, and genetic and epigenetic predispositions. Therefore, even as we look to the study of consciousness to help us comprehend pain, the converse—that the study of how we hurt might unlock the secrets of how we come to be—is perhaps much likelier.

The person most responsible for how we view pain today was neither a physician nor a biologist. French philosopher René Descartes (1596–1650) believed that only humans were capable of self-reflection, consciousness, and feeling pain. He considered animals no different from, say, a microwave oven sounding off alarms in response to electrical stimuli without any sense of what those alarms meant. An animal's cry was similar to your leg kicking when lightly tapped below the kneecap—a mindless reflex without rhyme or reason.

As hard as pain can be to tease out in people, given that the gold standard is still reliance on a person's description of the intensity of his or her discomfort, it is orders of magnitude harder to do the same thing in the slimy or crusty, slithering or lumbering, shaggy or slick animals that we inhabit this earth with.

Are animals conscious? And do animals feel pain? Answers to these questions can help us resolve a fundamental conundrum: Why does pain hurt?

When we observe single-celled bacteria, they clearly react negatively to uninviting stimuli. Yet because cells lack a nervous system, it is fair to say that they are likely neither conscious nor in pain.

At some point in evolutionary history, as organisms became increasingly complex, growing into multicellular invertebrates and then into vertebrate fish, they developed more sophisticated machinery with which to sense their environments. Navigating water is actually much simpler than traversing land, given that the array of challenges land-based animals experience can be far more diverse than what one might encounter underwater. Was pain one of those mechanisms developed to better survive our treacherous terrestrial environments, or were fish already capable of feeling pain before they evolved further?

We know that most invertebrates, such as sea urchins or starfish, carry some of the same nociceptive receptors we humans do to help detect threats, such as acid-sensing ion channels. Not only are these channels found in fish like trout and zebrafish, but they can also be suppressed with painkilling medications that work for humans. So, clearly these organisms experience nociception using some mechanisms similar to those of more evolved species.[8]

While nociception can easily be seen as a feature of intelligent life, how can one differentiate it from pain, particularly in fish and invertebrates? Such a question is enormously important: while there is general consensus that animals, particularly large mammals, feel pain at least on the same spectrum as we humans do, which has spurred growing attention to animal cruelty, many researchers argue that fish do not have the capacity to feel pain, and therefore their treatment need not conform to the standards we have set for mammals.[9]

Given that fish cannot simply tell us they are hurting, scientists closely monitor their behavior as they respond to noxious stimuli, picking up cues that indicate a key characteristic of pain: learning. An animal that experiences pain should demonstrate a change in its future behavior that goes beyond a simple reflex, a sign that the pain has left a mark in its memory. Ideally, this behavior should dissipate with the administration of painkillers. These "pain behaviors," if noted in fish, would mean not just that they experience pain but that they carry at least some form of consciousness. Evidence of pain in fish or invertebrates would also suggest that far from a recent development in biological history, pain, and therefore consciousness, might be an ancient aspect of life not requiring a very well-developed central nervous system to exist.

This is exactly what a spate of recent experiments appear to suggest.

There is a popular myth that a goldfish's memory goes back only about three seconds, and for this, goldfish catch a lot of flak. Yet one experiment showed that goldfish are far smarter and more self-reflective than their reputation suggests. Researchers delivered electric shocks to goldfish in only a specific part of an aquarium. After the shock, the goldfish began avoiding that part of the aquarium. In fact, as the voltage of the electric shock increased, so did the length of time the goldfish stayed away from the dangerous part of the tank.[10]

In a second experiment, fish were trained to feed in a specific part of the aquarium, where they would then be shocked. The fish reacted as they had earlier and zipped away. But then the researchers started tightening the screws, varying just how starved the fish were. They found that the more famished the fish, the more likely they were to return to the shock zone. This experiment showed that the fish remembered the hurt of the shock and its intensity and were smart enough to weigh the relative benefits of having food in the belly versus enduring the shock of the current, adjusting their behavior based on the price they were willing to pay. These experiments strongly suggest that fish feel pain and don't just experience nociception.[11]

Squids, which are invertebrates, are highly intelligent, and they too have been shown to learn from pain. And when that lesson is not learned, this can have serious consequences. In one experiment, researchers injured the arms of squid and then released into their vicinity their predators, black sea bass. Black sea bass have a propensity to target injured squids, but the injured squids were far more vigilant than uninjured squids and swam further away from the sea bass. Being in pain made the injured squids warier and more careful and, contrary to what one might presume, provided them a survival advantage over their uninjured peers.[12]

In the next step of the experiment, both the injured and uninjured squids were anesthetized with magnesium chloride, a natural painkiller that is activated in squids after injury. While this anesthesia had no impact on the uninjured squids, the injured squids failed to learn what their pain was desperately trying to teach them and were much more likely to be eaten by the sea bass than any other group. Far from helping, the painkiller actually cut the squids' lives short.

Despite the similarities that these experiments make clear, how fish and squids feel pain is quite different from how we humans do. When fish and squids get hurt, they appear to feel pain all over their bodies and are unable to pinpoint exactly where they are wounded. Inject a trout with acid, and it will start breathing heavily, shake its entire body, and rub up against the sides of the aquarium, regardless of where the corrosion is instilled. These behaviors reliably dissipate when trout are given opioid painkillers like morphine.

That fish and many invertebrates consistently demonstrate the ability to feel pain is the most compelling evidence that they also possess consciousness, representing at the very least an unfinished version of the painting fully realized in us humans.

Pain, however, is not an inevitable conclusion of nociception. In fact, there are moments when we experience nociception without pain. Henry Beecher, the first endowed professor of anesthesiology and a pioneer in bioethics, was in a field hospital during the Allied assault on Italy's Anzio beachhead during World War II, when he made a historic observation. The field hospital was receiving a stream of soldiers who had been "subjected to almost uninterrupted shell fire for weeks." Beecher interviewed soldiers who had just been brought in after sustaining extensive injuries, often from violent explosions, but were mentally clear, were not in shock, and had not received any painkillers for several hours. Only a quarter of these soldiers with brutal mutilations said they had enough pain to want *anything* done about it. And in the minority of soldiers who did report pain, there seemed to be no relationship between the extent of their injuries and how much they hurt. Beecher posited that the relief of not having to go back to the hellacious theater of war superseded whatever pain the soldiers might have otherwise experienced.[13]

You don't have to be on a battlefield to have nociception without pain. Athletes frequently experience nociception while playing a high-stakes game, allowing pain to enter only after the match ends. Yet to consistently experience nociception without pain is a recipe for a species' extinction. The hurt in pain is what helps us survive in an often brutal world—just ask the anesthetized squids who served themselves up on a platter for the sea bass. The transformation of the simple sensation of nociception into the complex experience of pain can easily be considered one of the greatest triumphs of evolution.

Pain exists and is so well preserved throughout species because pain is meaningful. It hurts because it is trying to teach us an important lesson—the biting sting of the Boston winds reminds me to bundle up in the winter to prevent my fingers from falling off. And because the longer an organism lives, the longer it has to hold on to its most distressing memories, the deeper the imprint of pain has to be in living things with long lifespans. As a child I remember sticking a pen in a power outlet and being enveloped in the momentary spine-tingling grip of electricity. I remember no other aspect of that encounter—not how old I was, where I was, or what I was thinking—only that fleeting feeling running throughout my entire body. That recollection is all I ever need to never approach a power outlet with anything pointy and to jump at the sight of my daughter attempting the same. Since humans live so long, we need to be especially tender to the hurt of pain; that memory might come in handy decades after the fact.

And there is ample evidence that what we find unsavory, as well as the threshold at which we do so and how powerfully we react to it, has been carefully programmed into our DNA. The more unpleasant pain is, the likelier we are to memorialize it and to do whatever possible to avoid it in the future.

There is always a reason why we all hurt the way we do.

Here's one illustration: While all fish experience pain, what individual species find painful can vary. Rainbow trout, for example, lack a nociceptive response to very low temperatures, whereas zebrafish don't. This is not a biological oversight but rather a deliberate design feature: rainbow trout live in cold waters, while zebrafish are tropical residents. The pain machinery of these fish is tailored to their habitats.[14]

And one reason why, unlike fish, we humans can pinpoint precisely where we hurt is that, unlike fish, which don't have hands or feet, we can use our dexterity to react to and minimize the extent of our injuries: we can clamp down on that bleeding wound on our shin, pull out that rusty nail buried in our heel, or slap away that hornet lodged in our shoulder.

I imagine the notion that pain helps us avoid danger is not a giant leap for most readers. But there is more to the story. When my daughter faceplants onto the floor, a nociceptive spike generates a painful experience in her mind. As soon as she falls, even before the first tears run down her cheeks or the booming bellows emerge from her itty-bitty frame, the very first thing she does is look around. The velocity of the tears, the amplitude of her cries, and the vigor of her protestation will all be determined by the presence or absence of her most vigilant and attentive caregivers. These mechanisms will be at play whether she is aware of them or not.

In a world infested with bloodthirsty predators and protective kin, the internal chronicle of pain has to be narrated to provide it a social dimension. As much as all pain is personal, to be fully realized what pain most needs is a witness. As much as we all believe that pain is a personal experience that can be nearly impossible to express in words, one of its chief functions is communication. As much as pain exists to be felt by the self, it lives to be seen by the other.

Few people look forward to a trip to the dentist: the chipping of the metal scalers, the high-pitched shriek of grinding enamel, the occasional stab of the dental drill in your throbbing gums, the taste of blood in the mouth that occasionally follows, while you sit stiff as an overdone breadstick, at the mercy of a stranger with two hands and half their face in your mouth. It can transform the most stoic of individuals into whimpering heaps. But what if your dentist weren't a stranger? Would things be different?

It so happens that my mom also was my orthodontist, and not coincidentally I happened to be her worst patient. With loud protests, frequent interruptions to gargle, and requests for more local anesthetic, my histrionics often left important procedures unfinished. I still have a crooked tooth she won't forgive me for.

To be clear, my mother was beloved by most of her patients. My behavior had more to do with whom I was around than with what I was actually experiencing. My protestations were amplified by our relationship rather than the instrumentation in my mouth.

When we look at the animal kingdom, pain and how we experience it appear to be fairly universal. From invertebrates to Homo sapiens, injury leads to nociception leads to pain. But how we react to pain, how we speak and act it into existence, has evolved much more recently.

The first behaviors that pain elicits are reflexes that protect the organism. The first and most important behavior is withdrawal. Often this is a spinal reflex that doesn't even need the brain involved to be carried out. In many cases, this will prevent any further injury. This withdrawal is then followed in many animals by rubbing or licking. These actions bring comfort and relief but also serve an additional purpose: they could remove any bugs, critters, or other injurious objects. Animals also limp if they damage a limb or will guard a wound so that the initial injury is not aggravated.

In addition to self-defense, pain behavior is an essential means of communication. There is no tractor beam more effective at pulling a parent toward a child than the latter's wailing. We see this in animals as well—mothers flock to their babies when they cry out in pain. And even as children become better at feigning agony, most parents know when the cry is simply an attention-grabbing act versus a real emergency.

Pain behaviors are a way to demonstrate how we feel to others.

And just like you might not share your credit card number or email password with strangers, animals are wary of whom they share their pain with. In one experiment, mice were injected with acid, causing them to writhe in agony. Yet how energetically the mice writhed depended on their company. If a mouse was injected with acid at the same time as one of its cage mates, and the two could see each other, they seemed to writhe more. Yet when the other mouse injected with acid was a stranger, they did not exaggerate their distress. This suggests that pain operates like a social contagion, a way of seeking help and revealing vulnerability, but only within one's circle of trust, lest it betray weakness to an adversary.[15]

Therefore, just as we modulate what we say in front of different people—we are much franker with those we trust than with those we don't know well—how we choose to act out painful behaviors can vary based on who our audience is.

The presence of a loved one can also be a potent form of analgesia. Mice are less distressed when a family member is close by. In the previously described experiment in which goldfish were zapped as they approached food, the fish were more likely to return to the shock trap sooner if another fish of their same species was in the vicinity.[16]

At times, pain behavior can even bridge communication between different species: a human infant's crying increases the level of the stress hormone cortisol not only in humans but in domesticated dogs too, showing that the ability to feel empathy can be far-reaching.[17]

A pain behavior doesn't have to be as loud as a cry to elicit attention: most mammals recognize a grimace as a sign of discomfort. And not all grimaces are the same: scientists have developed scales for different animals to infer, just from their faces, how much discomfort they are in.

Being able to recognize pain behaviors is critically important. They not only signal that a member of the clan needs help but may also indicate the presence of danger: the yelps of a bear stuck in a trap might alert other bears to be more vigilant. For intelligent animals, such signaling can be very instructive.

Not all pain communication works to an animal's advantage though: predators can easily identify prey in pain and preferentially hunt them down, as the sea bass did the injured squids. The cries of a hurt gazelle may draw as many cheetahs as earnest gazelles. Further, not all animals respond kindly to a peer in pain. Some are repelled by other animals in pain. Lambs have been known to attack others demonstrating pain, perhaps so that they don't attract wolves in the vicinity. Pain behaviors can also be damaging to the animals themselves: many animals in pain develop an aversion to eating, drinking, moving, and having sex, making them both less fit and also unattractive as potential mates.[18]

Pain behaviors are most nuanced and layered in human beings. While frequently referred to as a deeply private affair, pain is in fact a social experience. It forms a critical basis for how we reach out to or wall off from those around us. How and when we choose to demonstrate distress is variable, and that variation is mediated by situational context. Men, for example, are less likely to reveal discomfort in the presence of women, especially women who are more sexually attractive. Why? Men are generally wary of exposing vulnerability, fearing that it may be perceived as "unmanly." Even our withholding of pain behaviors is a deliberate act of social signaling, whether others know we are holding back or not.[19]

Our disclosure of pain is often a transactional act: we express it when we expect to be safely acknowledged. Those same men who act tough in front of women they don't know are much more likely to show exaggerated pain behaviors around a spouse or loved one, someone who represents their safe space and whom they expect care from. As people experience more pain, over time they start to withdraw into their safe space, avoiding unfamiliar folk, turning that safe space into a prison.

Our reaction to others' distress, even something as primal as a parent's response to their child crying, can vary widely. Almost half of parents respond to their infants' cries with thoughts of harming them. Unfortunately, parents sometimes act on those toxic urges; the incidence of shaken baby syndrome in fact rose during the COVID-19 pandemic. Yet how we react to a child's cry and what factors lead us from care to neglect to abuse can inform a lot about one of the most important aspects of pain: the response it elicits. The way we communicate our pain matters greatly here. If you tell a doctor that you are having excruciating pain in your knee and grade it as a ten out of ten, but you don't appear uncomfortable, are not limping, can walk on your two feet, and are not jumping out of your seat when your knee is touched or manipulated, the doctor is unlikely to believe you.[20]

There is much we can learn about ourselves and why we hurt from the study of the living beings we share our world with. We know today that we are in some ways similar to plants in how we detect pain, given that we share some of the same receptors. It is also clear that the ability to detect threats through nociception is essential for the survival and success of living organisms. And while nociception is an unconscious sensation triggered by potentially injurious forces, pain is an unpleasant experience that the conscious animal mind creates to help react to its environment, learn from its mistakes, alter its future behavior, and communicate with its friends and foes.

Yet the study of animals can only teach us so much about what ails human beings. Philosopher Julian Jaynes called animal research "bad poetry disguised as science." And if we only study that poetry, we might overlook the most dreadful dimension of pain, one that only we humans appear to fully bear. That dimension is suffering.[21]

Of the many peculiarities in the practice of medicine, one that is particularly peculiar is that doctors and nurses are highly unlikely to have lived with serious illness. Physicians mostly prescribe drugs, order imaging studies, and perform procedures they have never experienced themselves. They are like chefs who have never tasted their own food.

That changed recently, when my patient and I both got the same viral infection: shingles.

Even after one fully recovers from chickenpox, the virus that causes the disease, varicella-zoster, can remain dormant in our nerves, only to reactivate years later. Such a reactivation, characterized by a painful rash, is called shingles and is often precipitated by stress.

When a flesh-colored rash resembling bubble wrap appeared on the side of my torso, shingles was the last diagnosis I had in mind. I was much more suspicious that it was poison ivy, given that it popped up right after I had wandered into the leafy overgrowth in our backyard with my daughter. Yet the fact that it seemed to catch fire every time my shirt even grazed it made the diagnosis self-evident, given what I knew about shingles. However, after almost two weeks, a week after the bulbous eruption had begun to heal, the pain also receded.

As my rash faded and I began feeling better, my patient's trial was only beginning. Shingles can sometimes lead to a condition called postherpetic neuralgia, wherein people can have pain long after the shingles dissipates. Yet what he was experiencing was beyond anything I had seen even compared to other patients. Innumerable medicines and nerve blocks to quell the torment had failed to give him back a shred of his previous pain-free existence.

His pain by now had transfigured into something wholly different from what it was birthed as. Robbed of his dreams, he was so depressed he had essentially stopped eating. He showed me a picture of himself playing golf before shingles had crashed into him, and he looked unrecognizable. His agony was so consumptive, so ravenous, it had left him emaciated, sapped of all his strength and spirit.

At this point, we had already exhausted every conceivable medical intervention—there was nothing more we could honestly say or do to offer him relief. We didn't even know where his affliction really lay. Was the furnace in his nerves causing his catatonic depression, or was his depression keeping the flames in his nerves alight? Yet we were struggling to define something even more basic: Was he in pain, or was he suffering?

In my native language, Urdu, the word for journey—سفر—is a homonym of the word for suffer.

That coincidence could not be more apropos. Suffering is a voyage that one often sets out on after the onset of pain. Yet an injury is not a prerequisite for suffering. Losing a loved one, being fired from a cherished job, and finding out your favorite comedian is a sex offender are all events that cause us to suffer.

So what is suffering really? And how is it different from pain?

What pain is to the body, suffering is to the heart and soul. Pain derives from physical discomfort and suffering from mental anguish. This demarcation has arisen in Western thought from the works of René Descartes, who famously distinguished the mind and the body as separate systems.

This framework allowed us to differentiate ourselves from complex animals like fish and other mammals who feel pain but don't appear to suffer. Suffering has long been considered a uniquely human experience, often juxtaposed, on the other end of the emotive scale, with happiness. The world, as much as it seems to be a jostling ground between the forces of good and evil, is also very much a ring where happiness and suffering grapple with each other like sumo wrestlers in a dead heat.

The separation of the body and the mind propagated by Descartes allowed science in Western Europe to flourish because it granted biologists and physicians free rein in the physical world while allowing the church to maintain its hegemony over the metaphysical realm. Yet this historical demarcation represents modern pain science's original sin. Physicians are taught to think of disorders as either "organic," meaning they are real and have a physical origin, or "functional," a less direct way of saying they are "all in the patient's head." We doggedly fight diseases that can be seen under a microscope or palpated with the tips of one's fingertips. Yet, when it comes to disorders that cause suffering but cannot be detected by a lab test or CAT scan, we send the patient over to a therapist.

Most people understand on some level the different weight we ascribe to the realness of pain and the fuzziness of suffering. No wonder many will manifest their mental anguish and anxieties as physical symptoms such as abdominal pain or even seizures, because they know that in modern society something subjective might as well be unreal.

While pain, particularly chronic pain, can lead to lifelong suffering, where does one process end and the other begin? What is the difference between a normal response to pain, such as resting a gouty, inflamed foot, and the deep despair that often ensnares the victim of chronic pain?

Those who live with pain, who ruminate about it day and night, who can trace each vine of the poison ivy creeping up their bodies, might be best positioned to help tease out that distinction. One such person is Ann Marie Gaudon, who has lived most of her adult life with all dimensions of pain.

Ann Marie had a relatively normal life until after her first pregnancy. "One night I woke up with so much pelvic pain, I had no idea what was going on," she told me. "I took a cab to the ED where I was asked to give a urine sample, which was pure blood."

For the next few years, she kept going back to her doctors with symptoms resembling recurrent urinary tract infections, but her lab tests never revealed the presence of any bacteria. "The doctors wouldn't give me anything because they were waiting for a diagnosis and my pain went untreated," she said. "A chronic pain syndrome doesn't mean much to doctors. It's not like you have lymphoma." Lymphoma, Ann Marie said, would have been "legitimate" to her doctors.

Eventually Ann Marie was diagnosed with interstitial cystitis and bladder pain syndrome, but she never forgot the difficulty of not having her pain taken seriously until it was validated with a medical label. Ann Marie's own experience inspired her to help others in similar situations. She went back to university to train as a psychotherapist specializing in helping those struggling with chronic pain. While she can't treat "clean" biological pain, as she calls it, Ann Marie can help patients cope with their suffering, what she calls "dirty pain." "Clean pain is what we feel in our bodies, whether it's visceral, neuropathic or muscular pain. Dirty pain is how we react to our clean pain."

She describes suffering as coming in three progressive phases. "With mental scripts we are often finding reasons as to why we are in pain and we can be very harsh to ourselves," Ann Marie said. "We tell ourselves we are losers. That our life is over. And the more pain we are in, the more black that place will be." Our suffering then generates avoidance behaviors. "We might avoid exercise, we might avoid going out, avoid certain situations or meeting specific people," she added. Left unchecked, avoidance behaviors can lead to the final destination of suffering: the pain cage. "You are pretty much disconnected from the world. For most of us, engagement with other people, exercise and nature, are what give us joy. When those things are removed, that's when you are in real trouble."

Ann Marie reaches into her own catalog of experiences to help those she sees who are suffering. "I can't change the pain; my goal is to change their behavior. I am trying to change the relationship they have with their clean pain," she said.

Dirty pain comes with messy scars. "What inevitably happens is that people always don't come with just clean or dirty pain—they come with childhood trauma or other acute life stressors." So suffering begets more suffering, and people who have been dealt a toxic hand, either through traumatic life events, an inhospitable social environment, or an agonizing medical condition, are set up to go deeper and deeper into the labyrinth.

Human beings might suffer for reasons similar to why we have pain. Suffering, like happiness, is an adaptive and motivational force. Starving may cause you to suffer, which may lead you to hunt, forage, become full, and therefore feel happy. The more you suffer, the more driven you will be to find food, and the more successful you will be as an organism.

Suffering is essential not just for humans to survive but to find ourselves, to know who we are: not just nerves, synapses, and ganglia, not just bricks and mortar housing electrical currents that zig and zag and ebb and flow, but the I that emerges from it all.

A reasonable question might be why we have to suffer so much, a question that continues to elude evolutionary biologists. In *The Greatest Show on Earth*, Richard Dawkins wondered, "It remains a matter for interesting discussion why it has to be so...painful," going on to ask, like a devout believer, of the silent heavens, "What's wrong with a little red flag?"

We might especially balk at the contradiction of suffering in a society that continues to leap forward. Despite increased wealth and prosperity, scientific and medical advances, and an all-time low in interpersonal violence, human suffering, often hand in hand with pain, is surging like never before. And if current trends hold, its viselike grip on our species will continue to tighten.

If we are ever to have a chance of finding our way to a new understanding of these physical, emotional, and existential experiences, we must better delineate the differences between nociception, pain, and suffering. Even though it is their suffering that most people are concerned about, our training as physicians and nurses is increasingly detached from this reality. In an emergency room I worked in, patients were given iPads and asked to provide a stream-of-consciousness output of their pain on a scale from zero to ten. As I walked by their rooms, I would see people in hospital gowns propped on their beds, busy on their iPads, dabbing with their fingers, sending a running chyron to our computers. Here modern medicine had fundamentally failed the people it is meant to serve. It asked people to take the most complex experience they could ever have, one that fundamentally challenges the artificial distinction between the body and the mind, between the physical and the metaphysical, one that has emotional, spiritual, genetic, epigenetic, evolutionary, racial, and psychological dimensions, and reduce it to a single number on a ten-point scale.

The truth is that no two people experience the same pain, and no two pains experienced by the same person are similar either. Pain isn't just a sensation; it's an event in our biographies. Slap a hundred people, and you might get a hundred different responses. Strike one person a hundred times, and each strike might get a hundred differing reactions.

It is context not biology that dictates variation in pain. Pain from a fractured rib will feel different from pain due to lung cancer after a lifetime of smoking. A fracture can heal, but metastatic cancer changes how one views oneself and one's future. And pain from a tumor encroaching in the ribs will mean something different to a newly married young woman versus a Vietnam War veteran who still has posttraumatic stress disorder from his time in service.

In a landmark essay published in 1982 in the *New England Journal of Medicine*, Eric Cassell, a physician practicing in New York City, defined suffering as "a state of severe distress associated with events that threaten the intactness of the person." What adds suffering to pain is when people "feel out of control, when the pain is overwhelming, when the source of the pain is unknown, when the meaning of the pain is dire, or when the pain is chronic," he wrote. Suffering, there-fore, occurs when all the rules that we artificially assign to pain—that it should be meaningful, temporary, and diagnosable—break down.[22]

When people in pain come to a physician for comfort, they come with these rules lurking in their subconscious. More than just hoping for the pain to be killed, they often seek answers. Almost none of my patients have ever asked me how their cells hurt, which nerves are misfiring, what parts of the brain are pulsating. Almost all of them want to know why they suffer.

Yet, this is not something doctors are taught to probe. Doctors are largely trained like car mechanics: to perform annual inspections before the brakes wear out, to grease the wheels when they start squeaking, to replace the engine when it dies, and when nothing can be done, to send it to the scrap yard to be crushed into tin cans. Clinicians are deft at controlling nociception from a surgical incision or an inflamed appendix but are less adept at treating pain and are woefully inept at relieving suffering. We spend years rote-learning the complex routes arteries and veins take and the myriad steps biochemical reactions perform in our bodies. This information helps us achieve the needed grades on standardized tests but rarely, if ever, comes up in the real-world care of human beings.

Cassell, born in 1928, went to medical school in the 1950s. I caught him off guard when I called him unannounced in the middle of the day on his publicly listed telephone number. I didn't expect to reach him, and it was clear he initially wasn't ready to discuss something he had written about decades ago. But he quickly warmed up, launching a volley of anecdotes at me as I struggled to keep up. "I don't get to talk to people often," he told me with boyish keenness.

"When I graduated from medical school, I thought I should get out pretty soon, because we will cure everything and all the diseases would be gone. And yet all we had were antibiotics."

"I graduated into the world of certainty," he said. "When the patient didn't respond, it was never because we didn't do the right thing."

But the world that Cassell graduated into was different from that promised to him in medical school. While it was true that previously incurable diseases like gangrenous infections were now curable, he found that for all its bravado, the medical machine had done little to drain the river of despair he found himself swept up in as a physician. He was increasingly drawn to this elusive human condition that he had been taught nothing about as a student.

"Suffering is the entrance to the person," said Cassell, echoing the Persian poet Rumi, explaining why he spent his life exploring the concept. "It is the door to something much larger."

In the half century that separates his training from mine, instead of walking through that door and getting closer to the person, medicine has walked away. The experience of being a modern physician boils down to memorizing facts, mastering computer programs to perform clerical work, and performing procedures like plumbers or electricians. One mandatory test all physicians take to get their medical licenses is a video game that presents virtual patients and asks what tests and procedures to perform. Any wrong decision in this choose-your-adventure game causes it to abruptly move on to the next case without any explanation of what mistake was made. This is ironically emblematic of what it is like to be a modern-day clinician. Shortly after I had spoken to him, on September 24, 2021, Cassell passed away. His words, perhaps the last interview he ever gave, represent a gift, a baton he entrusted me to carry and pass on to future clinicians.

The medical profession teems with compassionate doctors and nurses, students and techs, sitters and aides, receptionists and administrators. Yet the sum is much weaker than the parts. By turning persons into patients and healers into providers, and by separating the body from the mind, physical sensations from emotional states, and pain from suffering, medicine is nothing more than a misguided miseducation in mortal misery.

The spheres of nociception, pain, and suffering both overlap and diverge. You can have pain without nociception, as is the case with patients who have phantom limb pain long after their arms or legs have been amputated. You can have nociception without pain, such as what a person in a coma might feel or a soldier might experience while running for their life unaware of the bullet lodged in their leg. You can even have pain without suffering, as evidenced by people who have a predilection for spicy food or sadomasochism. And, of course, we have plenty of capacity to suffer without any physical trauma ever inflicted.

Nociception is the most basic and most easily understood aspect of pain that all living beings possess. Pain connects us to other animals, providing clear hints about the conscious awareness that we appear to share. And it is our ability to suffer that distinguishes us as humans, opening us to a dimension of agony no other living being will thankfully ever know.

The last few decades have seen pain increasingly painted as a physical sensation. In essence, it has been conflated with nociception. This narrow view of pain has impaired doctors' and nurses' ability to separate nociception, pain, and suffering, doing a disservice to those who come to us for help. It is fair to say that pain provides a meaning to nociception, a meaning informed by our lives, our environments, and our cohabitants, with suffering the interpretation of pain. But by treating pain as essentially nociception, we have robbed people of what their pain means and the suffering that it manifests in its wake. Medicine's failure to effectively define nociception, pain, and suffering as overlapping but distinct entities that all combine to afflict the anguished is the reason most clinical interventions for pain that use painkillers or procedures and focus just on its most basic aspect fail: because they do little to address the multidimensional aspects of what it means to hurt.

A journey that reaches the true meaning of pain, incorporating the complete nature of suffering that spans mind and body, could help design a healthcare system that is truly "person oriented" as opposed to merely using that phrase as a buzzword on billboards. This journey to further our understanding of pain must move beyond a digit on a ten-point scale. To break the wheels of woe so many patients find themselves trapped under, we must master the actual biology of how our cells hurt, how they let the body know what they detect, and how the human brain turns nociception into pain and pain into suffering.

2.

HOW WE HURT

The Biology of Acute Pain

God whispers to us in our pleasures,
    speaks in our consciences, but
    shouts in our pains.
It is his megaphone to rouse a deaf
    world.

— C. S. Lewis

IT WAS THE KIND OF dreary Boston day that makes you long for spring. The wintry storm that started with snow in the morning had turned into rain by the afternoon, covering the town in a muddy-slushy amalgamation. I reached the medical office building that housed the Pain Management Center of Brigham and Women's Hospital, where I work, and took the elevator up to the third floor, where I met Rob Edwards, a pain psychologist and researcher.

Edwards led me down a windowless, claustrophobic hallway to an office where I would learn about how I hurt. In order to understand pain in all its wondrous and diverse glory, I had volunteered myself for an afternoon of pinpricks, pressure cuffs, heat pads, and ice-water baths. Edwards opened the door into a brightly lit room, and two researchers working in his lab, Sam and Marise, greeted me with warm smiles. Sam was a PhD student and pain researcher; Marise was the lab manager. They would be administering the quantitative sensory testing (QST) I was there for, which can tell what a person's pain threshold and pain tolerance are, so I could better understand the machinery that makes us miserable. They were about to become my tormentors.

"Pain threshold is when you first experience a painful sensation, when it turns from non-noxious to noxious," Sam told me. "Pain tolerance is a combination of your body's nociceptive system and your own psychological and social experience over time." While the pain threshold indicates when you first perceive something as painful, pain tolerance is when you act to remove yourself from that situation. "It's when you go from 'I can keep going' to 'I CAN'T KEEP GOING!'"

Sam and Marise started the testing by readying a set of cylinders with retractable blunt needles at their tips. The needles would poke the top of my middle finger as my hand lay palm down on a table. Each cylinder had a different weight, which would be applied in a stepwise fashion. "You can keep your eyes open if you would like to," Marise told me, asking me to rate my pain on a ten-point scale. I went from a 0 with the first two cylinders to a 1 with the third, which is when they stopped. This initial test was done to make sure it was okay for me to proceed.

Next they wanted to see how I would do with pain summation. Pain summation involves a noxious stimulus being delivered repetitively. In my case, Marise took the same cylinder with the retractable needle and dropped it on my finger rhythmically like a drumbeat. She asked me periodically to rate my pain. Over a few seconds it went from a 0 to a 0.5 to 0.75 and then a 1, which is when she stopped. This, I was informed, was a normal experience of pain. Temporal summation—increasing discomfort from a consistent repetitive noxious stimulus—is normal. It's a general index denoting how susceptible the nervous system is to facilitating pain signals. Chronic pain, for example, can cause a reorganization of the nervous system, and those living with chronic pain are likely to summate faster.

I looked back down on where Marise's instrument had been tapping my finger. It had left a small, pink mark that looked a bit like a spider bite. The mark burned dimly, and I could still feel the fading echo of the tapping. Throughout the test, I was somewhat conscious of not wanting to squirm too quickly. But my evaluators saw me staring at my hand.

"Do you notice any after sensation?"

"Umm...yeah."

"Is it painful at all? Our patients often describe it as painful."

It wasn't really pain, more the hazy shadow of pain, a memory that receded as the redness faded.

Next they inserted my thumb into an algometer, a pistol-shaped instrument with a clamp at the end of its barrel that would squeeze my thumb with increasing pressure. Marise ratcheted up the pressure slowly, instructing me to tell her when it turned painful. The feeling gradually rose, eventually crescendoing when the rest of my fingers went numb and started to tingle. It was almost as if the pain in my thumb had stolen the ability of my other fingers to feel anything. When I finally asked Marise to let go, the algometer left a gray dent on top of my thumb. Marise wrote down the reading on the instrument after every round of pressure, incredulous at finally testing someone with relatively normal pain thresholds. "Wow, we haven't had a healthy one in a long time."

Sam then placed a small plastic adaptor next to my arm, which would detect how sensitive I was to heat. It was connected to a laptop on which I could see the temperature rise. In my hand was a clicker, which I could use to cool down the instrument immediately. The heated plastic quickly went from warm to hot to *ohmygodmakeitstop*. For most people, there is not much difference between when the heat first becomes uncomfortable and when it becomes unbearable.

For the last event in this tetrathlon of torment, Sam and Marise had me dunk my hand in cold water. The initial rush of cold made me finally understand why they call it frostbite. With every second, the cold clutched my hand more firmly in its jaws, and just as it peaked at what I called a seven, Sam squeezed my right shoulder with the clamp of the algometer, drilling down into it. My brain felt like it was going to tear; I had only enough of it to tend to one of my simultaneous abuses. As soon as Marise asked me to let her know when to stop, I immediately asked for my release.

"This is when people stop liking us," said Marise with a laugh.

"The paradigm we tested here is conditioned pain modulation," said Sam. "Pain inhibits pain."

My results showed that my threshold for pain in my shoulder doubled when I had my hand simultaneously dunked in cold water.

Much tougher for me was the cold water. When I was told afterward that my hand had been in the water for just forty-five seconds, I was shocked. I would have bet it was several minutes. At its worst I called the pain a seven, but this was assuming I knew what an eight, nine, or ten would feel like. My hand was still part red and blue. Marise shocked me by revealing her personal record: six minutes.

"If you make it past the one-minute mark, your hand just becomes nice and numb," she told me. "Also, I had to beat the intern's time."

"I have been struck by the fact that people's perception of their pain sensitivity shows no correlation with what people actually achieve in the lab," Rob Edwards told me sitting in his office. Behind him were pictures of him and his young children and a "World's Greatest Boss" mug; on his desk was a calendar with a cigar-smoking bulldog, slumped in slumber, with the word "Swamped" on top.

Edwards had a boyish demeanor. Every time I asked him something, he would take a deep breath, his smile would widen, and he would tell me what a great question it was. He was wearing a purple-checked shirt with a tie. I could easily see him helping fix my laptop at an Apple Store. His green work badge was faded and peeling on the sides, suggesting there were many more miles on his wheels than met the eye. "We have got these big beefy guys who say they can tolerate pain better than most people. And they are the ones who often tell me, 'I can't believe they let you do that to people!'"

In research studies, QST has been used to predict how much pain a patient might have after an operation and therefore help the team plan how to manage it. However, QST is rarely used in routine clinical practice since few insurance companies cover the testing. One would think that a more personalized approach to pain management, tailored to where every individual falls on the very wide bell curve of pain sensitivity, would be routine, but it still remains largely experimental. So even as all of us experience pain differently, and pain itself can have many different causes and manifestations, our cookie-cutter approach to pain management remains blunt like a hammer, and everything looks like a nail.

In the last century, we have unraveled most deadly human diseases, from childhood infections to heart disease. We have used these insights to double our average lifespan from the forties to the eighties. Diseases that were once cataclysmic, like smallpox and bubonic plague, have become historical curiosities. HIV, only recently discovered, has already been transformed from a death sentence into a chronic condition. Even with all these changes, one thing has remained constant: pain.

The big difference between pain and all these conditions is that pain is not a disorder but a normal bodily function. While disease affects a subset of human society, pain is hardwired into every single member of it. Yet, despite all the advances we have made in overcoming pathology—eliminating inherited disease with gene therapy, transplanting hearts or building artificial ones, training our immune system to fight cancer—we are nowhere close to fully understanding our physiology. More so than anything else I have ever studied in the biological sciences, our understanding of pain is unsettled, with the tectonic plates constantly shifting beneath our feet.

"My guess is that the presumed squishiness of pain," said Edwards, "is a big reason why it's taken longer to reach that same level of understanding in the field of pain than the fields of cancer, heart disease or HIV."

While much remains shrouded in mystery, we know a lot more about pain today than we ever have before as we inch closer to its ultimate truth.

When my chronic pain was at its worst, its biology did not concern me. How it affected my life, how it hurt, how it trapped me in its cage, and how I could eventually escape it were my chief concerns. The mechanics of how forces in the world around me and the maladies festering within could turn my insides out didn't arouse my curiosity. Their outcome did.

While pain is multidimensional, my experience in the QST lab seemed to flatten it into its most simplistic components. Even such a personalized approach to assessing my fidgety nervous system took place in a completely artificial environment. Yet understanding the mechanics of acute pain is necessary if we are to gather not just what we know about it but how much still remains unknown and how limited our tools to assess and alleviate it are.

The most basic feature of nociception is the detection of aversive forces in our environment and transmission of these from our bodies to our brains. This transmission is, like many of the processes in our body, electrical, and it flows through our nervous system.

The nervous system has two basic elements: the central nervous system, made up of the brain and spinal cord, and the peripheral nervous system, which includes the numerous nerves that reach every distant bit of us. From the tips of our fingers to the parts of our back that we can't reach when they are itchy, our skin is wired with nerve endings that are always processing information, even when we may not be paying attention. Of all the things our nerves are watching out for, they are best designed to alert us to harm. Our nerves inform us when our joints are being twisted too hard, when our muscles are being stretched to shreds, and when our bellies are about to burst from the bottomless brunch buffet.

Our nerves can taste four flavors of harm: heat, cold, mechanical force, and chemical stimulation. Using specific receptors, our nervous system can convert these stimuli into different electrical signals.

Let's take one of these noxious sensations: heat. When heat receptors in our skin detect sufficiently high temperatures, they change shape and allow a calcium ion to enter, triggering an electrical signal to go up the nerve fiber toward the spinal cord. These heat receptors, called TRP and discovered by David Julius in 1997, the other scientist awarded the Nobel Prize in 2021, are also sensitive to chemicals like capsaicin and acid. Capsaicin adds the spice to spicy food. Spicy is an acquired taste, as no child is ever born with an affinity for it, and when we adapt to spicy taste, we acquire not so much a palate for flavor as a palate for pain. This same family of receptors makes our eyes sting and tear up when we are cutting onions or being bombarded with tear gas.

Acid is also detected by acid-sensing channels, which we share with other species such as trout. The ability to detect acid in the body is very important. Our body uses oxygen to work, what's called aerobic exercise, like strolling in the park. But if we exercise more vigorously than is comfortable, such as, say, when running up a steep hill, our body switches to anaerobic mechanisms. This causes lactic acid to build up in our muscles. Such acid buildup due to overzealous exertion, felt as muscle cramps, could be damaging to the body, so we need means to detect it. Acid can also build up when the body is deprived of oxygen. When heart muscle is starved of oxygen-rich blood during a heart attack, it begins to disintegrate, leading to the release of lactic acid. Nociceptive fibers sense the presence of this acid, which causes pain in the chest.

Pain receptors in the skin are more than just conduits of information. They protect us not just by generating pain but by sounding the sentinel alarm for our immune system to jump into action. Nerve endings in our skin do this by releasing a stew of inflammatory substances that cause the surrounding area to become red and swollen in response to noxious stimuli. This in turn both activates other nociceptive fibers in the vicinity and lowers the threshold at which they fire, making even previously innocuous stimuli painful. This phenomenon, known as neuroinflammation, leads to the release of substance P from the nerve endings, which causes surrounding blood vessels to leak, sending even more inflammatory mercenaries to the region in pain, causing the nociceptive umbra to grow further.

Two types of rails transmit these messages to the brain. A-delta nerves have large-diameter fibers that are covered in fatty myelin sheaths, lending them some electrical insulation. They can convey nociception quickly with little interruption. Pain from A-delta fibers can be traced to exactly where it originates in the body. Nociception is transmitted much more slowly through the more numerous un-myelinated, thin C fibers. The discomfort that C fibers convey is much more diffuse and unlocalized. The nociception from A-delta fibers is a pinpoint prick; from C fibers it is dull and pressure-like.

We used to believe that any nociceptive nerve can detect heat, cold, mechanical pressure, and chemical irritation. But newer findings seem to suggest otherwise. To get a sense of just how underdeveloped the foundation of pain science is, consider this: in August 2019, re-searchers from the Karolinska Institute in Sweden discovered an as yet unrecognized organ of cells lying underneath the skin of mice. This organ is comprised of cells that live in the deeper dermis layer of the skin, covering nerve endings in tentacle-like extensions that extend up to the epidermis, the outermost layer of the skin, forming a mesh that detects mechanical force. While almost all of our conventional understanding of pain detection has been centered on naked nerve fiber endings in the skin, the fact that these bare endings are in fact covered by a layer of cells has only just come to light, suggesting that there is a lot we don't know even about some of the most basic mechanisms of nociception.[1]

In an ingenious experiment, the researchers used light to activate this layer of cells in the footpads of mice. Not only did the light activation of these cells result in a spike in electrical nociceptive activity, but the mice responded with the telltale patterns of being in pain: they withdrew their affected legs, licked and shook them, and then guarded them after the light was turned off. The researchers proved conclusively that this layer of cells is central to the detection and transmission of nociception. Yet how it modulates nociception still awaits further study.

When noxious signals travel up the spinal cord, passing through the brain stem, and enter the brain, it is there that dispassionate nociception is infused with the full smorgasbord of emotions, memories, and expectations to become pain. New ways of measuring brain activity are opening up these processes as never before.

Given the complexity of pain, it is not surprising that there is no single dedicated pain center in the brain. It takes several different parts of the brain working in unison as a network to generate the experience, painted and layered with sentiment and expectation, and dictated by attention and recollection, that we casually refer to as pain. This network is called the neuromatrix of pain.

As nociceptive fibers emerge from the top of the spinal cord and cross the brain stem, they enter one of the most important parts of the brain: the thalamus. The thalamus, whose name derives from the Greek word for "the inner chamber," is a node of gray matter that serves as a relay station between the body and the forebrain (the part of the brain that includes the cerebral cortexes, the slimy, bulbous tiara on top of our brains) and holds the key to our sentient consciousness. With the exception of smell, every human sense has levers located in the tiny thalamus. For nociception in particular, the thalamus acts as a hub of sorts and is instrumental in transforming it into pain.[2]

Noxious stimuli entering the thalamus come along two major information pathways: the sensory-discriminative pathway and the affective-motivational pathway. The sensory-discriminative pathway is concerned with the basics of pain: When did it start, where is it, and how intense is it? The affective-motivational pathway informs us of the emotional aspects of pain—how unpleasant it makes us feel—and guides how we adjust our behaviors in response.

These two distinct aspects of pain can at times be hard to separate: how bad pain makes you feel has a lot to do with how deep the wound is. The hotter the handle of the frying pan, the more it will hurt and the faster you will let go of it. Next time, you'll use an oven mitt.

The distinction between these two aspects of pain can be a matter of life or death. One of the most frequent symptoms I treat as a cardiologist is chest pain. In fact, chest pain is one of the most common causes of emergency room visits in the United States. While many cases of chest pain are nonthreatening, a critical proportion of patients will be experiencing fatal catastrophes like heart attacks, torn aortas, and punctured or clotted lungs. The trouble is that for many patients, benign acid reflux and the early stages of a heart attack can feel very similar—both can feel like the chest is burning. Patients who are concerned about their hearts, who are aware of the symptoms of a heart attack, or who have had previous heart disease may rush to the hospital for potentially lifesaving treatments. Yet many people, women in particular, are more likely to delay coming to the hospital because they don't think of heart disease as something they might be at risk for. This is one important reason many women stay at home, waiting for the pain to go away, and pay with their lives. It is the fear of a heart attack, rather than the intensity of the pain, that causes people to rush to the emergency room for potentially lifesaving therapy. When the pain doesn't arouse dread, it can actually be even more dangerous.[3]

To untangle the physical and emotional aspects of pain, Canadian researchers interviewed a group of young, healthy volunteers who were responsive to hypnosis. The researchers submerged the left hands of these volunteers for sixty seconds in a water bath that was either lukewarm (35°C/95°F) or "painfully hot" (47°C/117°F). The researchers had the volunteers do this while lying in a PET scanner, which measures changes in blood flow to different parts of the brain;

when a part of the brain is activated, its thirst for oxygenated blood increases.[4]

After a baseline had been set, the researchers used hypnosis to suggest that the volunteers felt an increase in the intensity of the heat, without actually altering temperature of the water. While the subjects who had been told the water was increasing in temperature felt the pain more intensely, their PET scans showed a change in only certain parts of the neuromatrix of pain, while other parts didn't seem to be altered. The part of the brain that responded most clearly to this sensory-discriminative aspect of pain was the somatosensory cortex, which receives, processes, and integrates all of our senses to produce our conscious experience.

In a subsequent experiment, the researchers followed the protocol identically, but with one big difference: instead of the intensity of the heat, they suggested the hypnotized volunteers alter the perceived unpleasantness of how the heat made them feel, activating the affective-motivational pathway. This time a different part of the brain shone brightly on the PET scan, the cingulate and insular cortex, part of the enigmatic limbic system of the human brain. The researchers largely proved both that how intensely painful something is can be disconnected from how awful it makes us feel and that we have different parts of our brain modulating those two aspects.[5]

One reason pain has been so hard to study and understand is that most of our research has been based on animals. But the mechanisms that have made pain a raging pandemic and a defining feature of modern life are not the animalistic features of pain, which are inherently adaptive and provide us an edge in hostile environments with limited resources. In humans, these basic features of pain are housed in the somatosensory cortex, the part of the brain that works with computerlike organization. It is predictable and its jurisdiction has clear boundaries.

What has made pain more challenging to grasp and manage is the part of our consciousness doused in emotions and desires: the limbic system.

The human brain can crudely be divided into three overlapping organs, each representing a leap forward in evolutionary development. The tail of the brain, which includes the brain stem and basal ganglia, is considered the reptilian brain and helps us perform our most animalistic functions: running, eating, breathing, fighting, copulating. As reptiles evolved into mammals, they developed islands of brain tissue shaped in an arc, called the limbic system, which regulates emotions. And as mammals came to be, they developed the third organ, which sits on top: the cerebral cortexes, which tame this emotional inner beast with measured reason. All three are critical to our ability to feel, react to, learn from, and remember pain.

If we are to understand pain as a hyperspecialized defense mechanism that now causes more suffering than it prevents, it is critical for us to understand the role of the limbic system in shaping how we come to feel and live with pain. And of all the emotions that cohabitate with pain, none is more consequential than fear.

It's often hard to tell where fear ends and pain begins. Both protect us from harm, both ensure we don't do stupid things, and both push us to do everything within our means to rectify whatever led to our feeling the terror that pain or fear leaves in its wake.

And while we fear the dark, fear creepy crawlies, fear heights, there is one thing we have always feared more than all these horrors combined: death. This is the main thrust of terror management theory, derived from anthropologist Ernest Becker's Pulitzer Prize–winning book *The Denial of Death* (1973), which attempts to understand every human action as a reaction to our mortality. To defy our mortality and to overcome the schism that arises from our transient existence and our inability to reconcile with our ephemerality, terror management theory states, we created the greatest products of civilization: art, religion, culture, and literature, to name a few. We also sublimated our fear of death into medical research that has led to the average human lifespan doubling over the last hundred years. This dramatic and unprecedented extension of the human lifespan has never been replicated even for cells or bacteria in a laboratory.[6]

Yet, as we have successfully turned death from an unpredictable bolt of lightning into, in most cases, a predictable, age-dependent event, we have changed how we live. Instead of acute illnesses that come and go, most people now live with chronic conditions that can be managed but rarely cured. Even an infectious disease like COVID-19 is highly skewed in its lethality toward those of a certain age. And as death approaches, most people undergo procedures and surgeries, get poked with innumerable needles, and endure many uncomfortable manipulations and indignities, often ending with extreme maneuvers such as chest compressions or electric shocks when their hearts stop beating. This new face of modern death increasingly leads many people to say no to more procedures, tests, and hospitalizations when the end is nigh. They would rather let nature take its course than live in agony. For them, the fear of death has been usurped by the fear of pain.

Like pain, fear influences many of our decisions. Yet some people don't feel any fear at all. The most famous such person is a mother of three in her forties simply known as SM. Researchers have subjected SM to many cruel experiments. She has been made to visit some of the most haunted places on earth, shown some of the most horrifying horror films ever made, and handed live snakes and tarantulas. Life hasn't been any more charitable: she lives in a crime-ridden neighborhood, has been held at knifepoint, and has endured violent domestic abuse. These experiences have left her feeling a lot of things, except for the most obvious: fear.[7]

Instead of being scared by the "monsters" in the haunted sanatorium, she scared a few of the actors back. Instead of being repulsed by the venomous snakes and tarantulas, she wanted to pet them. Instead of panicking when a man held a knife to her throat during a mugging, she remained unflustered.

Not only can SM not feel fear herself, but she cannot identify fear in others either. She can't even draw a fearful face. The reason behind

SM's fearlessness is not some martial arts training or a unique life-affirming experience. SM has Urbach-Wiethe disease, a rare inherited disorder that leads to hardening of the skin and, in some cases, parts of the brain. In SM, the part of the brain responsible for inducing fear has become rocklike. That part is the amygdala, the almond-shaped collection of cells in the limbic system, one for each half of the brain. Any reader of this book who has begun to notice the predictable turns in this maze shouldn't be surprised that, in addition to fear, the amygdala is central to another core human experience: pain.

The amygdala was discovered by German physiologist Karl Burdach in the early nineteenth century. We know today that the amygdala is not just one structure but a complex of thirteen interconnected nuclei that receive information from all our sensory organs and help to integrate them with each other while also enmeshing them with an emotional and biographical catalog. It is the bridge between the stimuli in our environments and our responses to them and has been preserved by evolution even as other parts of the brain have transformed.[8]

Neuroscientists at Stanford performed an ingenious experiment published in 2019 to figure out the role of the amygdala in the animal in pain. They first developed a process in which mouse neurons would turn fluorescent when activated. To peek inside their brains, the scientists mounted miniature microscopes *inside* the mice's heads to follow neuronal activity by tracking changes in the calcium content of nerve cells. These mice were then subjected to all the usual experimental stimuli for pain, including heat, cold, and pinpricks. The researchers found what they called an ensemble of neurons in the amygdala that lit up after the mice were subjected to these noxious elements. Within this ensemble, activation increased as the uncomfortable element intensified. Not only did the neurons light up more brightly, but the behaviors the mice demonstrated in response to the pain also became increasingly distressed.[9]

Some of these neurons stayed lit for up to a week after they were first activated, and over time, ever lighter touch generated an equally vigorous nervous response, until the neurons would register even innocuous touch as painful—a phenomenon referred to as sensitization. Even when the mice were anticipating harm, like when the needle approached but never touched them, the amygdalar ensemble popped with activity. And while many of these neurons perked up even when the mice were subjected to nonpainful but aversive stimuli, such as a foul smell, a bitter taste, or a loud noise, a subset of the ensemble responded to nociception.

The scientists showed that when it comes to pain, the amygdala appeared to play a pivotal role. But pain is multifaceted, so the real question was which aspect of pain the amygdala was responsible for. The clues lay in how mice react to pain. When hurt, mice first withdraw from whatever is causing the sensation, like a pin, just as we might—this is a reflexive action, one that might not even need any involvement of the brain. The second way that mice respond is by licking their wound, guarding the affected part, or avoiding the wounding edge. This second set of behaviors has to do with the unpleasantness of pain.

To conclusively show what role the amygdala played in pain, the scientists flipped the switch, selectively shutting down the ensemble they had identified. The mice were let loose on a track that had an intensely hot pole and an intensely cold pole. The mice with the dysfunctional ensemble still withdrew from the poles, similarly to mice whose amygdalae were functioning normally, but they were now much more likely to visit the poles again rather than to learn from their past experiences. What this experiment showed was that while the amygdala is not responsible for the mice sensing a noxious stimulus and reflexively withdrawing from it, it is certainly responsible for making them dread the agony associated with it.

The amygdala gives an emotional grade to our experiences. Chest pain will elicit, appropriately, an anxiety-filled visit to the emergency room in someone who has previously had a heart attack, while someone without this traumatic memory might take an antacid and wait it out at home. The amygdala's regulation and mediation of emotion-related learning behaviors is sophisticated, but it isn't perfect. A forceful grab might rip back open an old psychic wound, regardless of whether the intent is protective or loving, in someone who has previously been grabbed in an abusive context.

The amygdala is unique in that it can both dampen and amplify pain. The amygdala is the reason why some types of stress can reduce how much one hurts, while others can make things worse. The same proteins can have a painkilling and a pain-amplifying effect in the amygdala.

In fact, how an individual's amygdala is wired with the rest of their brain can determine how that individual experiences pain. Researchers at the University of Reading in the United Kingdom tested the pain thresholds of thirty-seven healthy people, then recorded their thresholds again after they had been shown a picture evoking a negative emotion (such as a plane crash or a gun pointed right at them). Using functional MRI (fMRI) scanning, they found something interesting: people who experienced their pain as more unpleasant when associated with negative emotions had a stronger neuronal connection between their amygdala and the parts of the brain responsible for sensing pain. This suggests that what we sense is interconnected with what we feel, depending on how our brains are formed. In fact, even among the thirty-seven people tested, the researchers found lots of differences between individuals. While the vast majority found their pain worse after seeing the disturbing images, some experienced no change, and a few actually felt better.[10]

Fear is as essential to our existence as pain. Animals with damaged amygdalae keep walking into traps, keep getting themselves killed. When SM got mugged and was almost killed, instead of altering her path, she kept walking through the same park. The amygdala, it would seem, is like the voice in our head that warns us away from danger — and from pain.

Even so, there is more to pain than simply the shock that it speaks into existence. As scientists have developed more sophisticated tools to pry open the limbic system, the supposedly primitive part of our brain, they have found that there is a lot more that the pained body feels. Newer techniques have uncovered a part of the brain that is central to every sting we feel, every flame we graze, even more so than the amygdala. If we are to understand how pain hurts, and hurts so much, we have to understand the part of the brain that is essential to what it feels to be human: the insula.

The insula is the part of the brain that helps us understand what it feels like to be in our bodies, allowing our consciousness to fully possess them. Its name is derived from the Latin word for island, and the extent of its role is only now being uncovered. When it comes to pain, the insula might just be the bridge that repudiates centuries of mind-body dogma that continues to hold back our understanding of pain.

The insula is the closest thing we have to a pain center in our brain. Work by Irene Tracey, a neuroscientist at Cambridge University and one of the leading pain researchers in the world, has shown that hurting will activate many parts of the neuromatrix of pain. But one part is always at play: the posterior insula. The localization of the sensation to the posterior insula provides, she has said, "a biological benchmark for agony."[11]

The moment you trap your hand in the slamming door, nociceptors in your skin, the bones of your fingers, and the ligaments in your hand will fire signals from the tip of your extremity to the spinal cord, where they zip up to the brain. These signals take two largely parallel paths. One will go to the brain and the amygdala, the affective-motivational path, while the other will go to the posterior part of the insula. This sensory-motor pathway carries the hows, whens, and wheres of pain.

The insula punches above its weight, and yet it was barely noticed by the man who made the first comprehensive atlas of the human brain.

Wilder Graves Penfield was born in Spokane, Washington, in 1891 and died in Montreal, Canada, in 1976. By way of Princeton, Oxford, Johns Hopkins, Harvard, and Columbia, he studied medicine, trained as a neurosurgeon, and developed what became a revolutionary surgical procedure for patients with epilepsy.

People with epilepsy often experience an aura—a visual, olfactory, or other sensory hallucination—preceding their seizures. While many patients respond well to medications, some continue to seize despite taking antiepileptic drugs. Penfield was singularly focused on helping these patients. He developed a surgical procedure in which he would poke different parts of the brains of awake people with debilitating seizures. When he reached the part of the brain that evoked the aura, he would zap that part, thereby relieving many people of their intractable epilepsy.

Penfield realized through performing this procedure that different parts of the brain evoked different sensations and were responsible for distinct functions. Through his work, Penfield developed something even greater than what he had set out to achieve: a map of the human brain. Penfield managed to reproduce every sensation and every movement in every part of the body except for one—pain. Not only is the human brain devoid of any nociceptive receptors—amazingly this allows some people to, say, nonchalantly play the violin while undergoing neurosurgery—but Penfield couldn't find even a single part of the brain that could be poked into producing a pang. Penfield's work seemed to fully reject the idea of a neural pain center, a puppet master pulling chains and changing gears, causing agony and anguish with every turn of the wheel and every crank of the shaft.[12]

Still, half a century after Penfield's work, an epileptologist in France named Jean Isnard took up the attempt to find such a pain center. He set upon this path serendipitously, after taking care of Jacob (not his real name), a twenty-two-year-old nomad whose life had been ravaged by a rare and crushing form of epilepsy.[13]

The aura that preceded Jacob's seizures was not a peculiar smell, an unusual vision, or a pungent taste: it was sheer, unvarnished pain. About once a week pain would engulf the entire left side of his body. It started off as a tingling sensation, then morphed into a burning one, quickly escalating into intense pain that often caused him to scream; it then became more throbbing as it died down over the next few minutes. Over time, these attacks became more frequent; people assumed Jacob had gone insane. Not until one of his bouts was followed by a full-blown seizure was the question of epilepsy finally raised. A variety of seizure drugs was prescribed, but each one of them failed.

An MRI of Jacob's brain showed that a small section hidden deep within a side lobe had an abnormal appearance: it was the posterior insula. Jacob went to the operating room, and when surgeons found the posterior insula and stimulated it with an electrode, they did something Penfield never could despite decades of electrified intrusions: they caused pain. And as they dialed up the current, Jacob felt it all: the tingling turning into burning turning into excruciating agony. The surgeons burned the right posterior insula to a crisp. Jacob has never had a painful seizure since.

Is the posterior insula really the mysterious pain center scientists have been looking for all these years? To find out, the same French surgeons and researchers dug deeper. They looked at video recordings and responses of 164 patients undergoing epilepsy surgery at their medical center whose brains were prodded with electric probes more than four thousand times. In just about 1 percent of all of these pokes, patients produced the telltale signs of physical pain: they grimaced, cried, shouted, turned pale, or turned red.[14]

Not only were all of these painful responses elicited by prodding the posterior insula, but each millimeter of the posterior insula appeared to be responsible for sensing or re-creating pain in a corresponding part of the body.

"Exploration of the insular cortex as part of the pre-surgical evaluation revealed the unexpected richness of the functional valencies of the insula," Isnard wrote to me in an email. "The insular cortex constitutes a true sensory hub...In this sense, we can say that the posterior insular cortex plays the role of the 'pain center' in the brain."

The ascent of pain as it zips through the spine doesn't end when it simultaneously reaches the amygdala and the posterior insula. When we feel physical discomfort, both its sensory and emotive aspects are merged. It is impossible to separate them because the pain that we experience is not bits and pieces of nervous code but an integrated experience that transcends much of how we define individual sensations, perceptions, or emotions. And this occurs because after these parallel pain signals reach their respective targets—the emotive system in the amygdala and the perceptual system in the posterior insula—they converge in the hub of conscious self-awareness in the brain: the anterior insula.

When you compare the blueprints of the human brain to those of our closest relatives, chimpanzees, there is one major difference: our anterior insula is disproportionately larger. Humans differ from apes in how we process our emotions, and that process is modulated by the anterior insula. Over time, we have come to understand that the mammalian limbic system is not some prehistoric relic of our crazed irrational past but an essential part of how we commune with our bodies and how we balance our urges and our actions. It is in the anterior insula where we interface not only with ourselves but with the world around us.

When you get cold while driving in your car, you reach for the dial to turn up the heat. When your stomach is grumbling because you have been fasting all day, you can't stop thinking about the leftover Halloween candy you could be wolfing down. When you see your child singing a song she has been practicing, your chest swells with pride, and time slows to a crawl. Whenever you have a desire that you quench or smother, whenever you have an emotion in response to the ecosystem you inhabit, it is your insula that tailors your senses to their social and moral milieus and thereby modulates your responses to your insides and outsides. The insula is important, then, because it grants us self-ownership, without which we have no agency. And without agency, we have no us. The insula has innumerable connections not just to the rest of the brain but to the body as well. Stimulation of the insula can either lower or raise your heart rate and blood pressure. And when we attend to our bodies more carefully, such as when we become consciously aware of how fast our heart beats, there is increased activity in the anterior insula. In this way, the insula is essential to our capacity for self-awareness.[15]

The insula is where René Descartes's mind-body dichotomy finally dies, for it is in the insula that our so-called mind and body become one. To perform its many actions, the insula contains large, specialized cells called von Economo neurons, which are only present in humans, apes, whales, and some other large animals. While no one yet understands exactly what these neurons do, they appear to have an important role in forwarding the anterior insula's agenda: transforming bodily sensations into introspective dives and transcendent flights.

Integration of signals in the anterior insula allows us to give meaning to bodily sensations, generating what we call emotions. The anterior insula is also where nociception, a fairly simply physical stimulus, transforms into pain, a highly complex hybrid phenomenon. The same scientists who found the two parallel pathways reaching the posterior insula and the amygdala, carrying the sensory and affective faces of pain, respectively, also showed that these threads go on to meet and become braided in the anterior insula.[16]

Tracking this journey all the way back to the insula makes clear that pain resembles an emotion generated by a physical sensation, an emotion that can exist even in the absence of nociception. If emotions were tracked on two main axes—valence and arousal, with valence denoting whether it makes us feel good or bad and arousal showing how forcefully it prompts a response—pain would be the emotion eliciting the deepest of negative valences and the greatest of arousals.

Sigmund Freud, before he birthed psychoanalysis, was a bona fide neuroscience researcher. After he enrolled in medical school as a seventeen-year-old in Vienna, his first instinct, naturally, was to study the sex lives of eels, but he was later distracted by the lab of Ernst Brücke. In his six years working at the Physiological Institute in Vienna, the mecca of neuroscientific discovery in the world at the time, Freud made some key discoveries, including that electricity travels around the body through a series of individual cells called neurons rather than through contiguous channels. He also started to lay down the foundations for the psychoanalytic theories that would fundamentally alter how we have come to think of the human mind.[17]

Here's what Freud thought about what happened to the body in pain: Sensations such as pain transferred elemental energy from our environments into specialized neurons on the lookout for them in our skin and organs. When noxious signals from these peripheral neurons reached the brain, they released an aversive chemical that unleashed a cascade of hormones, which produced emotions. Crucially, if the mere memory of a painful experience was recalled, the same chemicals and hormones would gush out, reproducing the pain. However, if this memory did not reach a certain evocative intensity, it would only be experienced by the subconscious. This theory forms the core of Freud's subsequent ideas on how we process trauma, how it touches us in ways both silent and deafening, and how it might lead to conditions such as "hysteria."

And yet, what Freud did next was perhaps most telling. After having finished writing *Entwurf einer Psychologie* (Project for Scientific Psychology) in 1895, instead of disseminating it, he buried it. He shifted his attention from the lab to the couch, focusing on psychology rather than biology. Freud's neurological theory of human psychology didn't come to light until sixteen years after his death in 1954. He had no intention of it ever seeing the light of day.

Why did Freud give up on neuroscience as a means to understand the true nature of pain? "Every attempt to discover a localization of mental processes has miscarried completely," Freud wrote in his book *The Unconscious*. "The same fate would await any theory that attempted to recognize the anatomical position of the system (consciousness)—as being in the cortex, and to localize the unconscious processes in the subcortical parts of the brain. There is a hiatus which at present cannot be filled…Our psychical topography has for the present nothing to do with anatomy." Freud felt that the science of neuroanatomy had yet to be refined enough to truly explain how the brain worked. The only instruments biologists had to explore the origin of pain back then were sharp scalpels and cleavers. Since then, while progress has been made, a central question remains: Are we any closer to the truth about the nature of pain?

For many years scientists have privileged physical sensations over emotional states, which is part of why labeling pain as an emotion may feel delegitimizing for some. But science is coming to finally understand how important emotions are to our survival. Perhaps Freud was on to something when he abandoned neuroscience, finding its tools too primitive to reach the inner recesses of our consciousness. His decision to focus on the person in the patient, the mind in the body, and the emotion in the sensation put him light-years ahead of others of his time.

In the decades since Freud, our understanding of pain has moved forward in tremendous ways. Just a few years ago, we could only see what function a part of the brain performed by cutting it off with a blade or zapping it with an electrode. The science of pain was based entirely on study of animals who couldn't be asked what they felt, forcing us to rely exclusively on interpreting their behaviors and attempting to analogize them to our own.

We have come a long way since, and at this pace an end to this scientific sprint might well be in sight. Optogenetics, which allows scientists to turn nerve cells on and off with the click of a button, has allowed us to discover an entire organ of cells, hiding in plain sight, responsible for detecting noxious stimuli in our skins. One can only imagine what such advanced techniques will teach us next about how we experience pain.

And yet the science of pain is riddled with many of the gaping holes that have marred science at large. Many important experiments and case studies in pain science have evaded replication and confirmation. Take SM, for example, the woman with the rocklike amygdala who never feels fear. Other research groups have studied patients with similarly damaged amygdalae, and while these subjects can't detect fear in others, they certainly feel it themselves. There is concern about whether SM's responses reveal more about what she wants the researchers to hear than about what she might feel herself. Similarly, while many researchers have pinpointed the posterior insula as the part of the brain most consistently activated by pain, some groups have found that the posterior insula reacts similarly to both noxious and non-noxious signals. Other research has fared even more poorly. A 2019 paper in *Science* showed that shutting down the insula took away the lessons painful stimuli taught mice. Yet, only months after its publication, the lab failed to replicate its own experiments. It turned out that the researcher who had led the experiment had fudged the data, and that study has now been retracted.[18]

While technology has expanded our understanding of the brain immensely, it still has far to go. Take fMRI scans, for example, which feature prominently in many studies on pain and have been the cornerstone of neuroscience for the past few decades. While much better than the techniques that preceded them, they are woefully inadequate to obtain a complete sense of the human mind. Contrary to what you might think, fMRIs don't detect neuronal activity. They actually assess changes in blood flow to different parts of the brain, which can considerably lag behind the electrical activity, barely keeping up with the lightning-fast signals that zip around our nerves. The fMRI can't even tell whether the signals in the brain are going from top to bottom or bottom to top. They also have limited resolution: a hundred thousand neurons can exist in one pixel of the fMRI image.[19]

The various pathways that contribute to our experience of pain also make the sensation incredibly difficult to pin down. Even as imaging techniques have helped highlight the role played by the amygdala, thalamus, insula, and cortexes, no single part of the brain is specialized to detect and react to nociception, unlike, say, with vision. Components of the limbic system like the insula and the amygdala are central not just to pain but to a host of other emotions.[20]

And finally, what most experiments are measuring is not pain but experimentally induced nociception. It would be unethical to inflict real pain in humans, especially the sort of debilitating discomfort that I see in the hospital or the life-altering agony that I experienced myself.

While undergoing the QST testing, I was fully in control of my pain, I could make it stop whenever I wanted, I was in a safe environment, I knew my body was not in danger, I was surrounded by empathetic concerned people, and I knew why I was uncomfortable. In the real world, people in pain are often not in control, often can't make the pain stop, are often in unsafe environments, often aren't sure if their bodies are secure, are rarely surrounded by concerned individuals, and all too often don't know why they hurt. The people who volunteer for research studies are mostly young and healthy without any serious medical conditions and are frequently studied as they lie perfectly still in fMRI scanners. During the QST testing, even though I hurt, I never suffered.

Yet, despite the optimal conditions of the QST testing environment, I was still quickly reaching the end of my will and tapping to surrender despite my desire to not appear weak or feeble. When pain came knocking, no matter what the circumstances, I caved to it again and again.

And as close as we may be to pinning down where pain lives in the brain, the fact is that so far most of our energy has been focused on the simplest form of pain: the one you experience when you walk into a glass door or when you play basketball on the weekend only to learn you are not a teenager anymore. This pain is physiologic and essential for our body's continued upkeep. Yet, in our contemporary world, when we speak of an epidemic of pain, we aren't referring to more people getting hurt falling out of trees; we're concerned about the emergence of a different form of pain. While acute pain is very much part of the normal functioning of our bodies, for many people, pain has become so distant and detached from its earlier form that it has mutated into something else altogether.

Pain has transformed from a symptom into a disease.

Just as the body works hard to maintain homeostasis, a state of equilibrium in its physiologic processes, the mind always seeks to achieve psychological neutrality. Nothing is more disruptive to that sense of balance than pain, and nothing can derail our lives more than one of the great scourges of our time: chronic pain.

3.

NO END IN SIGHT

How Chronic Pain Erases a Person

Life hurts a lot more than death. At
the point of death, the pain is
over.

—Jim Morrison

A S WE MAKE OUR WAY through our lives and the present
recedes into the past, pain often helps us decide what to commit to memory and what to let fade. I am famously fickle in my
recollections. While this is bad for many obvious reasons, making, for
example, the intense memorization required through medical school
particularly arduous, it does have some advantages. For one, I can
barely hold a grudge. Even in the middle of disagreements with my
wife, I often forget what we were originally arguing about. When I
think back further, my childhood seems like one long happy blur,
shaped more by what my parents have told me since than flashes I can
confidently own.

The only thing that makes my memories stick is pain.

In one of my earliest recollections, I am on top of a slide
during school recess. I am probably five or six years old. I remember the scene being chaotic, with kids running in all directions
across the dusty school grounds. The slide rises up like a pyramid
in the midst. There is intense competition for the fleeting whizz
down its slope. I am one of those jostling on the small platform
on top. Until I fall.

Was I pushed, or did I slip? I have vacillated on this ever since I woke up in the school infirmary afterward. What I do remember for sure is looking up at the spinning sky, cloudless and icy blue, without even being granted the time to be scared, right before the lights went out.

I am not alone in conserving these memories of pain.

I met Kyle (not his real name) as he was waking up from anesthesia, having just undergone a cardiac catheterization procedure. Most people who undergo this procedure need a light sedative to make them a bit sleepy. Some require a squirt of local anesthetic.

Not only did Kyle need general anesthesia that put him in a coma, but half a dozen nurses had to pin him to the bed as he woke up thrashing violently. He was an athlete, one of the strongest boys I had ever seen, and it took many bodies to corral him—all because of how much pain he was in.

Kyle had received a heart transplant years before, and its expiry date had arrived. On average, a transplanted heart lasts just over a decade before it starts to peter out. His transplanted heart could barely beat now, and he would not live any longer if he didn't receive another transplant.

Kyle had a lot to live for. A superstar at his high school, with a dazzling smile and an effortless charm, he had already come further than anyone would have believed possible. He had beat the bad heart he was born with and, perhaps even more importantly, the broken home he had grown up in. Yet, when I spoke to him, he was very clear: he would rather die than get another heart.

It took me a while to gain Kyle's trust. We bonded over the Xbox he had installed in his hospital room. I shared a copy of a game that he wanted to play. All these years around doctors and nurses had actually drawn him to the medical sciences. He wanted detailed explanations of what was going on in his body.

And it was only after he trusted me that he shared why he never wanted to go under the knife again. Ten years before, in the middle of his first surgery, Kyle had woken up. Lying on that flat gurney under the blinding lights in the operating room, he could hear and feel everything but couldn't move a muscle. The medical team had no idea. Imprisoned in his body, Kyle felt the scalpel cut into his neck. He wanted to scream, wanted to tell them to stop. He lost consciousness not from the anesthesia but from the pain overwhelming him. Ten years later, that memory still haunted him.

For most of us, pain is thankfully a temporary guest—like a fly that enters the house when the door is left ajar. It zips around, banging into the glass windows, until it receives a permanent respite, usually from the underside of a slipper. Even if not smacked, a fly lives for only a couple of weeks before it dies. Short-lived pain is something we need to feel so that we can live, unlike that bewildered housefly.

Many now live with more than just a flashback of pain. One out of every five people alive today lives with chronic pain. For those in pain chronically, their house is always full of flies, buzzing furiously in all corners, at all times, slamming into walls, drowning out the sunlight with their swarm, going nowhere because all the doors are locked, the windows bolted.

Acute and chronic pain have long been considered to be on the same spectrum, the only distinction being their duration. New research paints a completely different picture: the only thing chronic pain seems to share with acute pain is that it hurts.

Few people have been as influential in their area of expertise as John Bonica, who created the field of pain medicine almost single-handedly. Born in 1917 on an island off the coast of Sicily, Bonica was first drawn to medicine when he saw his mother, a nurse, assisting in the lancing of a breast abscess. Bonica fainted at the gruesome sight, but the experience nonetheless left him with a desire to become a physician. His family immigrated to the United States when he was eleven years old. His father had left his job as a deputy mayor in Sicily to work as a laborer in Brooklyn but died only four years after they moved. Thrust into the role of breadwinner at the age of fifteen, he hawked newspapers, shined shoes, and sold groceries to support his family.[1]

While in school, Bonica took up amateur wrestling, fighting professionally through college and medical school. On his way back in the trolley from one of his bouts, he met Emma, a Venetian girl who had come to watch the match. So smitten was Bonica that he promised to marry her, and after many years, he did.

After training as an anesthesiologist, Bonica joined the army and was sent to Washington State to work in one of the largest military hospitals on the West Coast. Pain infused Bonica's every waking moment. He saw thousands of soldiers who had suffered horrific injuries during World War II. Many of them remained in pain long after their scars had healed. As if this weren't enough, the love of his life, Emma, almost died while in labor with their firstborn after she was given poorly administered anesthesia. All the while, he continued wrestling anonymously as the "Masked Marvel." But this alter ego broke his body apart: he had four spine surgeries and countless operations on almost every major joint in his body. Ultimately, these personal experiences became the basis from which he would begin to develop the clinical specialty of pain medicine.

One of Bonica's great contributions was the medical textbook *The Management of Pain*, a fifteen-hundred-page tome published in 1953. It was the first book in any language solely dedicated to alleviating pain, and it standardized how pain was treated. (To this day, long after Bonica's death, it is an essential presence in the office of any pain specialist I know.) Bonica perfected and popularized the use of epidural anesthesia during childbirth, in which anesthetic is injected directly into the spinal cord of a person in labor, allowing them to be awake yet free of discomfort. When Bonica had his second child with Emma, she became one of the first women in the world to deliver after receiving an epidural. Bonica, alongside a nurse and a neurosurgeon, also put together the first multidisciplinary pain center, emphasizing the many-sided nature of pain. Bonica was really one of the first people in modern medicine to take seriously the notion that there is no one magic bullet for pain, that not all pain is the same, and that helping those who suffer requires a team approach focused on the entire person. He founded the International Association for the Study of Pain, launched the first clinical training program in pain medicine, formalized the first universally accepted definition of pain, and led the leading scientific journal *Pain*. He is the most influential pain doctor to have ever lived.[2]

Bonica's work focused particularly on pathologic rather than physiologic pain, a pain that to him "in its late phases, when it becomes intractable...no longer serves a useful purpose and then becomes, through its mental and physical effects, a destructive force." By the time he published the second edition of his textbook on pain in 1990, now about two thousand pages in length, Bonica seemed to have identified which aspect of this pathologic pain he found abnormal. And he now had a new name for it, defining chronic pain as "pain persisting beyond the usual healing course of an acute injury or disease, or the pain recurring at intervals for months or years."[3]

Bonica had come to think of pain as either acute pain in the wake of an injury or disease, whose fate was tied to its precipitant, or this chronic pain, which seemed to live on long beyond the initial insult. There could have been other ways to classify pain—say, based on the mechanism of the pain, or where it was felt, or how much it affected the person who felt it—but to him there was something unique about this zombified pain that lived on long after it was supposed to have died away.

Bonica was far from alone in coming to this conclusion, and the 1990s would mark an inflection point as chronic pain moved from the shadows and into the limelight with the introduction of the blockbuster opioid OxyContin and the ascent of a movement labeling pain as the "fifth vital sign." Often unaccompanied by gaping wounds or bleeding lacerations, chronic pain had been inherently invisible. Suddenly, it was everywhere to be seen.

Yet John Bonica never won his own personal battle with chronic pain. He often needed strong painkillers just to function and "probably had more nerve blocks and trigger-point injections than anyone else on the planet," wrote a biographer. He would find momentary peace only when floating weightlessly in a swimming pool or in the sea when he traveled back to his birthplace in Sicily. Perhaps the only thing that kept him alive was his enduring love for his wife. When Emma passed away in July 1994, John Bonica died the very next month.[4]

For much of human history, life was short and exciting: death could strike out of the blue anytime, anywhere, anyhow. Injuries, infections, and malnutrition bagged most human bodies. But as public health and hygiene and modern medicine gained steam, they made our lives not only much longer but much more predictable. Death is now very strongly tied to age, and premature deaths, though highly visible and tragic, are a drop in the mortal bucket. Heart disease and cancer account for two out of three deaths around the world even in the midst of a worldwide pandemic of historic proportions.

Now that we live longer lives, we have greatly increased the number of disorders we carry with us. As life expectancy has increased, so have the years we spend in poor health. While there is some variation depending on how the question is asked and how chronic pain is defined, most studies indicate that pain is a persistent presence for one in five Americans, which comes to sixty-six million people. Some studies estimate that the number is as high as one in three. And the number of Americans often troubled by pain continues to rise.[5]

The burden of chronic pain doesn't fall evenly. Chronic pain is a multiplicative disadvantage, much more common in groups who already face inequity: women, racial and ethnic minorities, those who are poor or poorly educated, and those with other medical conditions are all at much higher risk of experiencing chronic pain.[6]

Some have suggested that chronic pain is a uniquely American syndrome. In a survey of about fifty-two thousand people, 34 percent of Americans reported that within the last month they often or very often had bodily aches or pains, compared to an average of 20 percent of people in other countries. At least in this survey, Americans were far and away leading the pack when it came to persistent pain. However, other studies refute the notion that Americans are more likely to have chronic pain compared to those living in comparable economies. The "third world" doesn't fare much better. An analysis of developing countries shows that, much as in higher-income nations, about one in five among the general population experiences chronic pain. Chronic pain, therefore, is a truly worldwide vexation.[7]

What is behind all this pain?

The most common cause of pain around the world is the one I live with: low back pain. Low back pain is the leading cause of disability in the United States, accounting for more than three million years lived with disability. The rest of the world isn't faring much better: low back pain was found to be the leading cause of years lived with disability in both men and women in a study of 195 countries and territories. Not only that, but the number of years lived with disability from low back pain went up 30 percent from 1990 to 2007 and a further 18 percent from 2007 to 2017, to about sixty-five million years globally.[8]

While low back pain is the undisputed heavyweight champion of chronic pain, there are several close runners-up, including headaches, especially migraines. In the United States, migraines are the fifth most common cause of years lived with disability. Globally headaches affect three billion people, are the second most common condition causing disability, and account for the second most years lived with disability. They are closely followed by neck pain and conditions like osteoarthritis and rheumatoid arthritis, which are also some of the most common causes of disability around the world.[9]

These conditions, most of which are grouped together as musculo-skeletal disorders, aren't just painful. They are also extremely costly.

The United States spends more money treating musculoskeletal disorders than any other condition. In 2016, $380 billion was spent managing musculoskeletal disorders, more than the amount spent on heart disease and cancer combined. Back and neck pain alone were responsible for $135 billion. Unlike with other conditions that get a lot more attention, most of this spending is for middle-aged people rather than those who are older or approaching the end of life. And spending on musculoskeletal disorders is growing by 5 percent a year, compared to a 1 percent annual increase for cancer and a 0.5 percent annual reduction for heart disease.[10]

But numbers tell only part of the story. Chronic pain can impact human life in ways unlike any other disease. It is often said that what doesn't kill us makes us stronger. Chronic pain is this belief's strongest refutation.

Disease is a primal part of every human being's story, a rite of passage we all have to undergo as we slowly move between birth and death. And the way we often make sense of our diseases is through storytelling. Patients tell a tale, and it is the job of the nurses, physicians, social workers, receptionists, transporters, case managers, and myriad others who together form the faceless amoeboid mass that we call the healthcare system to listen.

Of course, the interplay between patients and the healthcare system they are enmeshed in is not one-sided. Medical professionals help assemble the narrative. Trained over years of school, with simulated actors, or through multiple-choice questions predicated on whodunit-style clinical scenarios, clinicians come to expect a certain cadence to patients' presentations.

You don't even have to be sick, or caring for someone sick, to recognize that classic arc of illness: The vibrant person who precedes the patient is struck by a mysterious illness, often but not always manifesting as physical discomfort. The now patient, often positioned as a fighter, charges forth in lockstep with their medical team in search of not just the putative cure but, even more importantly, the reason they were paradropped into this minefield to begin with. This spirited and pugnacious quest for a diagnosis is the plot of most medical narratives. Despite its amorality, the illness is not the villain, for when identified, even if not conquered, it provides meaning and resonance to the person's biography, building a new identity far more formidable than that of the naive, vibrant person of yesteryear.

Chronic pain, however, does not conform to these rules and stereotypes. Chronic pain affects people in ways that almost no other ailment does. And while it affects people in myriad ways, its most deadly feature is that it disrupts the way a person moves through their life, the narrative they define themselves by, the arc of their stories.

Drew Leder was a medical student at Yale when he decided to take an unusual hiatus. He went to get a PhD in philosophy. His main area of focus was phenomenology, the study of conscious experiences. "I became interested in the philosophy of the body, especially the experience of pain and illness and dysfunction and how it was treated by the medical system," he told me. Although he returned from his hiatus and finished medical school, Leder found what he was looking for in philosophy and never practiced medicine.

As for so many others interested in chronic pain, Leder's interest was chiseled into him by his own long history with it. "It's a little frowned upon in phenomenology to stick too close to one's own experience since you may fail to capture the full cultural spectrum of what people go through," he said, but added, "I don't think I would have been able to interrogate pain if I hadn't experienced it myself."

A healthy body feels absent and invisible. As we go about our lives, we focus on the world around us rather than the one within, which seems to work autonomously. "Often when we attend to our body it is during times when we feel pain or distress," says Leder. "Because the body is naturally absent when functionally well, by forcing it upon our attention during times of dysfunction, it dys-appears to us."

That the body becomes so present during sickness creates a wedge between it and its occupant. People in pain feel like their own body is an adversary. Their bodies alienate them from their healthy absent-bodied pasts, and when it comes to chronic pain, they rob them of their futures as well.

"People with acute pain can see the horizon of health and normality that they will be able to return to," said Leder. "The pain of childbirth, or the pain of appendicitis, one has this feeling that this will be over. Chronic pain has this existential dimension—what if this never goes away? What if it tarnishes and blackens every month or every year?" With a past that has become unfamiliar and a future shrouded in dread, people with chronic pain become trapped in a never-ending present.

Leder's research led him to study a group of people whose experiences mirrored those of patients with persistent pain: people in prison. It wasn't until I spoke to Leder that I realized how much my own experience of chronic pain mirrored incarceration.

I never knew what my back did for me until I broke it. Turns out, it did everything. It helped me stand, sit, and lie down. It helped me walk and run. After I hurt it, an MRI revealed that long before my injury, my spine had been deformed by years of poor posture and poorly performed exercise, which had straightened it like a poker out of its natural S-shaped curvature. And now a disc was bulging into my spinal cord, sending chills down to its tail all the way to my toes. My entire body became tense, a rubber band stretched to its wavering limit.

My expansive life was now reduced to my dorm room, barely bigger than a bathroom. Sitting in a car could be agonizing. A staircase loomed like an impassable wall. It hurt so much to walk to the common bathroom that I often peed in the sink in the room. At my worst, I couldn't even get out of bed, even though it hurt so much just to lie there.

My physical shackles also locked me out of my social life. If friends weren't kind enough to come to my room and take pity on my pathetic existence, I would never be able to see them. I ran short of friends quickly.

Even as small as my room was, I couldn't attend to it since my back commanded all my attention, all the time. It not only trapped me in a claustrophobic physical space but also jailed me in the one point in time I wanted nothing to do with—the now. Pain prolonged every second I lived, making every micro-decision arduous, making every day feel like an eternity. As much as I wanted an escape from my agony, I remained locked in place as the pain sapped every joy I could ever experience.

Everything about my past life seemed so far removed, I felt another person had been living it. The things that gave me joy—the basketball court, the gym, the track—now only bred resentment. I couldn't write because my imagination couldn't not regard the pain. It was so ceaseless that I remembered being pain-free about as well as I could remember floating in my mother's womb. When I looked to my future, I saw myself wading up an endless river of white-water woe thrashing me mercilessly.

All I had was where I didn't want to be. I was locked up in the moment, in the penitentiary of the present.

Like prison, chronic pain can take a person's community away from them. Many patients attempt to rectify that loss of social support by seeking medical help. "Sometimes people come to the health system looking for that, but are likely to be disappointed," said Leder. "The insurance companies don't reimburse for emotional support. It can leave someone very unheard."

Finding a diagnosis for chronic pain is the only way to get one's sentence cut short. While a diagnosis might help with treatment, to the person in distress it can provide something even more coveted: meaning. And yet the nature of chronic pain means that far from being an ally, for many people, the healthcare system becomes as much an antagonist as their ailment.

We may have entered the age of big data, but to understand the experience of those living in pain, the gold standard remains good old-fashioned qualitative research. To canvass what we know about what chronic pain has wrought upon people, the National Institute for Health Research (NIHR) in the United Kingdom funded a meta-ethnography, a collective analysis of what patients with musculo-skeletal pain go through. The researchers screened more than three hundred studies, selecting seventy-seven to synthesize the report. At more than two hundred pages, the report is an essential dissection of this ailment and the people it afflicts.[11]

The researchers identified five themes that define the struggles of patients with chronic pain. The first two—the fights to affirm oneself and to reconstruct oneself in time—were a direct function of this dis-ruptive disease. People with chronic pain are contending with a body gone rogue and threatening to evaporate their identity. The disease fractures their sense of time, leaving them paralyzed in the moment, unable to plan for the future or be spontaneous.

What's more devastating is that the next three struggles the research-ers identified—constructing an explanation for suffering, negotiating with the healthcare system, and proving legitimacy—are all toxic side effects of modern medicine, the clinicians it has trained, and the healthcare systems it has propped up. Far from providing relief, the medical system can leave many with chronic pain worse off than they were before.

There is no force that pushes us toward introspection quite like pain. The person in pain, as I know from experience, is hypervigilant, focusing on every twist their body takes and every surface their body touches. Such hyperawareness can be incredibly taxing, and it can often misfire, causing one to ruminate on every ache, every twinge.

The struggle to find an explanation for suffering is a direct arti-fact of the practice of medicine, in which meaning comes from a diagnosis. A diagnosis opens all sorts of doors for patients, reflecting how a doctor's scrawl on a pad or some hastily typed words in the computer can impact someone's entire life. It gives them hope for permanent liberation rather than ephemeral relief. It lets them feel like they have a physical disease rather than a mental condition, that what they have is "real" rather than in their heads. When they stand in front of an X-ray machine or are stretchered into the doughnut-shaped void of an MRI machine, almost every patient with chronic pain hopes something will light up, that something broken will be found. The last thing they want to be told is that everything looks good.

How we treat pain, how we view human suffering, has changed considerably since the end of the nineteenth century. Modern science changed human life so rapidly that it gave people "future shock." One would think that seismic shift would reverberate nowhere more powerfully than in the body of the man or woman in agony. Yet, when it comes to how a pained person is treated by the modern healthcare system, far from progression, there has been a regression, reflected most directly in the fourth theme the NIHR researchers highlighted: how people with chronic pain struggle to negotiate the healthcare system.

When Lara Birk's right leg collapsed in the middle of a soccer game, she initially thought she'd developed shin splints; she'd been running a lot that summer, training to play for the varsity squad as a junior. But the pain seemed to be out of proportion to that. No one at the field could tell what was going on, and eventually she was sent to the emergency room.

"The doctor kept telling me to stop being a crybaby. He kept asking my dad questions and wouldn't even make eye contact with me," she told me. "Another doctor told my mother it was all in my head and that she needed to take me to a psychiatrist."

Birk struggled for another day and a half in the hospital before someone finally figured out what was going on: she had acute ex-ertional compartment syndrome, a rare condition in which pressure builds up in the muscular portion of the arm or the leg. However, when her medical team placed a pressure gauge on her right leg, the pressure appeared to be under control, though she kept having pain. Eventually they realized that the gauge was not in the right place. As soon as they measured the actual pressure in her leg, she was taken for emergency surgery to relieve the stress. If the diagnosis had been delayed even a few hours, her surgeons told her, they would have had to chop her leg off.

While a diagnosis saved her limb, Birk became something I wish on none of my patients: the medically interesting case. Doctors were constantly filing in and out of her room to look at her leg. "They hushed me when I spoke so that they could talk to one another as they pointed to the exposed tendon, palpated the lump of leftover muscle, and poked their pocket scalpels into the necrotic flesh," she wrote in an autoethnography.[12]

This was just the start of Birk's journey with pain. Now in her mid-forties, she was in the hospital for six weeks after the operation and still using a wheelchair at the time of her discharge; she would walk with crutches for four years. The wound on her leg was ten inches long and four inches wide. Then her left leg developed compartment syndrome as well. She's had a total of fifteen surgeries, and even though the original "organic cause" of her pain apparently resolved, she has continued to be in distress.

When I spoke to her, it was apparent that the words her surgeons voiced hurt her even more than the incisions they made. "As a young girl, I wasn't taken seriously," she said. "I was often told I was being hysterical—that I was making it worse by paying attention to it."

Birk's erasure was relentless: "I would tell doctors to not touch this area but they disregarded it and hurt me."

Yet, because Birk's pain kept returning, she had no choice but to keep going back to the very clinicians who abused her. "The people I was going back to, I was invisible to them," she told me. "It was like gaslighting—I began to doubt my own thoughts. Maybe they are right and I am making this up. I internalized it and I am still working to undo that."

As heartbreaking as it is, Birk's tale is not exceptional. It is how modern medicine treats anything it doesn't comprehend. If doctors didn't learn about it in medical school or cannot make it go away, it must not be real.

Birk learned to negotiate the hegemonic structures that had come to govern her life. For the medical system, it is not enough for you to be sick; you have to act the part. "Over time I became practiced in what details to give when, and how long to talk when I walked into a room and met a physician for the first time," Birk said.

Birk is a proud person who never wanted to be obviously disabled by her pain, and yet she found that unless she playacted as she was expected to, people would not take her seriously. She didn't want to walk with a cane but would be heckled for parking in a handicap spot when she tried to brave it out without one. This core social function of chronic pain put her in a bind: underperform and you aren't taken seriously; overperform and you become suspect.

To prevent clinicians from hurting her physically, Birk became more likely to have someone accompany her to appointments and to be more assertive about not wanting to be examined on a first appointment. Being older and advancing in life didn't hurt either. "When [a doctor] found out that I was a dean at a college, his facial expression changed," she said. "He treated me very differently after that."

In other words, Birk took control of her story, something doctors are loathe to give away. She started to believe her own story rather than the one doctors were telling her. But she is especially well equipped to do that. Her experience of being in pain led her to train as a sociologist. And as she got better at performing her role, something else happened: "Just the act of telling a story can be therapeutic, can change its arc."

The average American doctor takes twelve seconds to interrupt a patient as they begin to tell their tale. Everything that follows is on the doctor's terms, in their chosen syntax. And Birk, as a white, highly educated, upper-middle-class person, has enough insight to know things could have been worse, writing that her "advanced social status" could blind her "to the many ways in which race and class can compound and complicate the effects of disability."[13]

Patients with chronic pain—unaligned with an algorithmic medical approach that prizes disorders it can visualize, characterize, and pulverize—have become pariahs. As disadvantaged as patients with primary mental health conditions like depression or schizophrenia are, people with chronic pain are even worse off because they exist in a purgatory between physical and psychological disease. This is the main reason why they experience the fifth and final major struggle identified by the NIHR research: the struggle for legitimacy.

The yearning for legitimacy can be all-consuming. It can annihilate a person's reality and, given time, eat into their entire surrounding world—starting with the people they love most.

Martha Lundgren had been married to her husband, David, for a decade before he developed a peculiar disability.

David could no longer sit.

Like so many people affected by chronic pain, Martha couldn't pinpoint the moment David's ordeal began. She remembered her husband having low back pain, but it was never too severe, until it suddenly was. The presumption was that David's sciatic nerve, which emerges from the spinal cord and goes all the way down the leg, was being compressed by a muscle in the buttocks known as the piriformis, which it passes through. Multiple surgeries by expert surgeons only made things worse. "It's the rarest of rarest situations where he sits more than twenty minutes at a time," she told me, "and usually that's the time it takes for him to go to the doctor."

"We used to be on the go all the time. We lived in Texas, twenty minutes from Austin. We loved going to hike and bike the trails," she said. "Now it's completely the opposite."

Being able to sit is a core human function, and I didn't fully realize that until I spoke to Martha. For example, David now stands when he goes to a restaurant. "There are some phrases, like have a seat or let's sit down, which are now emotionally painful," Martha told me. "I try not to use them."

As David is unable to sit in a car, let alone drive one, the couple bought a Subaru SUV and installed an air mattress in the back. David is strapped down on the mattress with chains to prevent him from sliding. The car is low enough to allow people to see him shackled in the back, and kids occasionally wave at him. One time, someone called the cops because they thought Martha was trying to kidnap him.

"My husband is an incredibly intelligent man, who had an enormous amount of responsibility in his professional life," Martha said. Before the pain, David had worked as a general manager at a cottonseed processing plant. Yet his condition hasn't just affected his frame. "It's that much harder for him to use his intelligence the way he used to. Instead of being able to fix this computer in 3 days, we are now at week number four."

Chronic pain has changed Martha's life as much as it has David's.

Often when two people meet, especially when they are young, they are healthy and don't have much of a medical record. Being healthy increases people's chances of being able to get married to begin with. Given time, though, we collect nicks and zits, sprains and sores, and, eventually, malignant tumors and failing organs.[14]

When Martha met David, he could sit. When he retired from his job, she had dreams of easing back from her job as well and focusing on things she enjoyed more. They had a large social network through their families, work, and church. Just as they were preparing for this life of leisurely comfort, piriformis syndrome burst onto the scene.

Martha is today far more dependent on her work, and not just because she is now the sole breadwinner. "We would be able to adapt to a lower income but not to not having my health insurance," she told me. "If we used Medicare, we would immediately fall into the donut hole, and our out-of-pocket costs would go up by $12,000 a year." Martha enjoys her work as a support manager in a healthcare

IT company, but her relationship with her job, as well as perhaps her marriage, has changed: "When you feel tied to something, the frustrations seem to get heightened."

Martha also does all the physical work at home now. "I am a farm kid, I grew up doing things," she said, "but I don't have a partner in that anymore. And sometimes I get really angry about that."

All of us occupy specific roles in the lives of our cohabitants. We are deployed across certain vectors of responsibility: at home, for example, while my wife cooks, I do the dishes. While she gets our daughter ready for school in the morning, I help prep her at night. She pays the bills; I do the taxes.

If I am in pain and I have to forsake my role, I am certain my wife will pitch in. Over time, though, unless I can offload other responsibilities from her, she will grow frustrated and resentful, while I will sink deeper into self-pity and self-loathing.

Caregiving for a spouse or loved one can be a daunting burden regardless of what illness they have. "When I watch other people in long-term caregiving situations, I know that long road is fraught with danger," said Martha. Some of those dangers are subtle: Martha used to exercise with David, but with his handicap she has lost her prime motivator, and her own health and well-being have suffered.

"People saw us as a couple. But it's different now."

There are many challenges unique to caring for people in pain. Caregivers need to be more forceful advocates because clinicians can often be at odds with patients with chronic pain. The lack of an explicit diagnosis to explain ongoing discomfort can cause friction between caregivers and the person needing care. But perhaps more than anything else, the chronic part of pain can become most toxic.

"I try to help, but I can't fix it," she said. "There is just no end in sight."

Pain is a social emotion, one that needs to be performed to be recognized.

The goal of a pain behavior, such as limping after stepping on a sharp piece of Lego, in addition to preventing reinjury is to communicate the pain to onlookers, translating this invisible and personal experience into a language others can understand. Patients with chronic pain have to calibrate this output carefully. In the absence of any obvious new injury, already struggling to prove genuineness, they have to play their hand just right. Overplay it, and you are the boy who cried wolf. Underplay it, and you suffer in silence.

Even to those who know chronic pain sufferers the best—their spouses—pain is accurately communicated only half the time. Spouses who underestimate their partner's pain might be critical or doubtful. Spouses who overestimate it, on the other hand, might feel even more protective, even more stressed, while pushing the partner deeper into their role as the pained person.[15]

Chronic pain can also be contagious. Spouses of those with chronic pain are far more likely to develop it themselves. One reason might be that caring for a person with chronic pain involves a lot of manual labor. Many caregivers, like Martha, sustain injuries, including torn knees and sprained backs, as they help their loved ones get out of bed, shower, or climb the stairs. The more pain they feel, the worse they become at fulfilling their other responsibilities, and the more brittle they become.[16]

Given all this, it will come as no surprise that there is evidence that the onset of a chronic illness increases the risk of divorce. But here is the real kicker: in heterosexual couples, the risk of divorce increases if the wife gets a chronic illness but not if the husband does. Women are much more likely to stick it out after their spouse becomes disabled than men. This has created another invisible epidemic: the majority of informal caregivers are women.[17]

Seeing a partner suffer has a measurable effect on your body. One study found that when spouses of people with chronic pain watch their partners carry heavy logs, their heart rate and blood pressure jump, compared to when they see a stranger perform the same task. These spikes in heart rate and blood pressure are even sharper when they simply talk about their spouses' pain and suffering. What's more, these changes are associated with an increased risk of heart disease.[18]

Chronic pain can come to define a relationship, and perhaps nothing is more dangerous than a coping behavior many patients with chronic pain turn to—catastrophizing—which can drag someone with chronic pain into a woeful vortex that often becomes the death knell of a relationship.

Tall and genial, Bob Jamison was wearing a pink button-down shirt when I met him at Brigham and Women's Hospital's Pain Management Center. His office was littered with ancient, likely nonfunctioning analog recording devices. There was a giant potted plant in the corner. "The day it touches the ceiling I am gonna retire," he told me with a laugh. Bob had probably told this joke to others before, but at this point the top of the plant was a hair's breadth away from making contact with the ceiling.

The specialty of pain medicine has grown considerably since John Bonica founded it. While he initially envisioned it as multidisciplinary, actual clinical training in pain medicine is heavily focused on procedures and drugs. Bob is not particularly proficient in either of those, and yet the patients he sees are the ones with the most intractable pain. "By the time I see them, patients have had chronic pain for three to five years and have seen ten to fifteen specialists."

Bob, a professor at Harvard Medical School, is a pain psychologist.

"Some people are not interested in seeing me at all. They still want to find the specialist who can find out what's wrong with them," he told me. "Sadly, medicine cannot always make things disappear."

Having taken care of innumerable people with chronic pain, I know exactly what Bob is saying. Referring a patient with chronic pain to a psychologist may appear to be an act of medical surrender— as though we have given up on finding the overzealous nerve or the crumbling bone that explains why someone continues to hurt. I see patients taking offense at the notion that their pain is as much a physical as a psychological disorder. A patient may see a referral to a psychologist from a medical doctor as a flagrant betrayal. Clinicians themselves may view sending a patient with chronic pain for a psychological referral as an abdication of their responsibility. Yet Bob doesn't see his work as at odds with that of people like me.

"I am a counselor for people who are suffering. I help people come to terms with how they feel and gain a semblance of control," he told me. "If your pancreas stops working, you have to make huge changes—you have to start checking your blood sugar, you have to start taking insulin. Chronic pain is the same way."

While psychologists like Bob can help almost any patient with a serious discomforting condition, their support is perhaps most necessary for those who catastrophize. Just as chronic pain is an exaggerated, enduring form of physical pain, catastrophizing is its emotional mimic.

Our understanding of catastrophizing was laid down by the American psychiatrist Aaron Beck, the pioneer of modern cognitive therapy. His work focused on how conditions like anxiety and depression stem from distortions that take hold of our thinking. These cognitive distortions can come in many forms for someone in agony. One might experience an improvement in their discomfort after exercising but might fixate on not being entirely pain-free, which would be an example of mental filtering. One might see a condescending physician and then assume all doctors are disdainful, an example of overgeneralization. One might feel that they deserve the disorder— say, arthritis—that causes their chronic pain, an example of personalization. Yet the cognitive distortion most damaging to the person in pain is catastrophizing.

Pain catastrophizing entails thoughts and behaviors that amplify the threat and seriousness of actual or anticipated pain. It has three main components: magnification ("I am afraid the pain will get worse"), rumination ("I cannot seem to keep it out of my mind"), and helplessness ("I worry all the time about whether the pain will end").

One of the earliest analyses of pain catastrophizing took place in the 1970s, when about a hundred undergrads in Ottawa were enrolled in a research study testing their cold-pain tolerance. "Catastrophizers" were people who said they became fearful before their hand was in the cold-water bath, who kept thinking, "I can't stand this much longer," or who likened what they felt in their cold hand to, for example, what it must feel like to drown in the frigid, winter sea. The study found that catastrophizers were much less likely to benefit from hypnotic treatment to help improve the pain since they just couldn't uncouple from it.[19]

This initial report has since been replicated innumerable times. Catastrophizing can occur in any painful situation—be the pain acute, as after a whiplash injury, or chronic, as in patients with fibromyalgia—and can predict how well patients will benefit from treatments, as well as which patients will have greater pain, disability, and opioid use in the future. This mental disruption also lays the foundation for future anxiety and depression. While the relationship between catastrophizing and pain has a chicken-or-egg quality, catastrophizing being strongly associated with the intensity of pain, most research shows that the impact of these magnifying behaviors is independent of how intensely the pain is felt.[20]

We catastrophize for the same reason that a calf bleats to catch her mother's ear after she trips and falls. It is our way of speaking the unspeakable into existence. Catastrophizing represents the socialization of pain, according to Michael Sullivan, a Canadian researcher who developed a widely used scale to measure the degrees of catastrophic expression. People who catastrophize display facial and vocal pain behaviors for longer when others are around than when they are by themselves. Theirs is a communal approach to coping, a literal cry for help.[21]

The brain often learns to catastrophize in the wake of trauma. This trauma can stem from a fractured leg or a dental procedure or, worse, from sexual abuse or growing up homeless.

But while catastrophizing may be an effective adaptation strategy for transitory acute torment, for the person whose pain is protracted, it can become a recipe for self-destruction, a sickness in itself. Intense pain elicits intense worry about the future and a sense of paralysis in the present, which only intensifies future pain. It has an oppressive, totalizing presence that often leaves our emotional selves unable to mount a proportionate response.

This cycle is untenable not just for the person in pain but for their entire care network.

To interrogate how catastrophizing affects relationships, researchers studied 144 couples, interviewing them three times a day for three weeks. One member of each couple had chronic knee pain from arthritis. They found that if patients catastrophized about their knees during the morning, their spouses exhibited greater degrees of depression, anger, and frustration later that day. The following day, instead of making empathetic gestures, their spouses were more likely to get irritated with them or to ignore their concerns. These punishing behaviors, in turn, further predicted even greater catastrophizing. Another research study suggested that men and women do not respond similarly to catastrophizing: men appear to develop more depressive symptoms when their wives catastrophize, but no such effect is noted in women whose husbands do the same.[22]

For someone in persistent pain, a yearning for kindness and love can paradoxically doom a relationship, invalidating a person's anguish in their own home and alienating the ones they depend on the most. And the couples who are most vulnerable to this corrosive cycle of vulnerability, hostility, and withdrawal are not the ones who sleep in separate bedrooms or live parallel lives. The closer the members of a couple are, the likelier they are to share the contagion of chronic pain.[23]

This is why the work of people like Bob Jamison is so crucial. "People [in pain] think of the worst scenario. My job is to prevent people from focusing on how things could get worse," he told me, "giving people a sense of control over how they feel and training them to catch themselves when they have worried thoughts."

One would think that listening intently to people's fears, speaking kindly to them in response, and teaching them ways to gain autonomy over their bodies and minds would not be the purview of any one particular medical specialty. All physicians, especially those who treat patients with the invisible, unglamorous, and unsympathetic syndrome of chronic pain, should be able to provide comfort, even if they don't have a cure.

Yet these qualities, though perhaps as effective at pain relief as the strongest narcotic, are in short supply, according to Bob. "We teach people all the knowledge—understanding and diagnosing problems—but we don't teach them how to be a caring physician," he said. "A lot of people went into pain medicine because there was a lot of money to be made. People are getting into this field for the wrong reasons."

"We do a lot of implanted devices and a lot of injections, which helps pay for my service, but much of the staff is focused on billing," he told me. "We do procedures but we don't fix anything."

Bob doesn't just indict pain specialists; he calls out every part of our healthcare system. Our inability to develop what are often facetiously called the "touchy-feely" bits of caring for patients—you know, those that put the care in healthcare—has led to a widening abyss between the healthcare system and the people who come to it for help. The reality of medicine looks nothing like what's in the brochure.

Yet there is another layer to how chronic pain can affect a person. Unlike many other conditions, such as heart disease or cancer, which mostly affect older people, the vast majority of patients with chronic pain are middle-aged and in the prime of their productive potential. Chronic pain can not only derail people's careers but impact their ability to work long after its intensity has abated.

Deep into my medical training, twelve years after I entered medical school, something quite unusual happened. I fell in love.

I hadn't always loved being a doctor. Medical school had been a long exercise in disillusionment and anomie for the first two years. I struggled to learn about how cells work, which nerves go where, and why our hearts look the way they do.

Everything changed when I started the clinical phase of medical school starting in my third year. I enjoyed every single thing I did: I wanted to be a psychiatrist, then a pediatrician, then an ICU doctor, and then a cardiologist, depending on which rotation I was doing. So long as I was taking care of people rather than buried in textbooks, I was happy. I had fallen in love with medicine, and it was during this upswing that I broke my back.

There is never a good time to get a life-altering injury. I had been despondent about my choice to be a physician, and right at the moment when everything seemed to come together and I started looking forward to the future, my dreams turned to dust. Instead of being in the operating room assisting in surgeries, I was taking off early and lying in my room, watching the ceiling fan spin and spin. I spent more time in the physical therapy suite than in the lecture halls. I resisted every impulse to look to the future because I was afraid of what I would see.

Eventually, even though I lost the sheen of invincibility so many young people seem to embody, I got better and was able to continue medical school, followed by research, followed by internal medicine training, until I started my fellowship in cardiology. Every time I started a new position, I had to fill out an employee health form. When asked to list any medical conditions, while there was always a box for back pain, I always left it empty. I wasn't taking any medications and had my pain under control with back exercises and improved posture. I felt like I had truly left it behind, rendered it from a pressing reality to a hazy memory. And with my pain fully in my rearview mirror, one day on my way back from work in the second year of my cardiology fellowship, nine years after my back injury, I fell in love all over again.

Interventional cardiology revolves around minimally invasive procedures performed in a suite called the cardiac catheterization lab. Performing a procedure was a meditative sojourn from the chaos of usual medical work. Instead of having to divvy up my attention between a dozen or more patients in the hospital ward or a series of increasingly short clinic visits squeezed one after another, I could just be fully present for the one patient in front of me.

Walking back from work while I was working in the cath lab, despite the losses and fumbles, I was filled with a sense that what I did was meaningful. It was the closest I had come to having a spiritual experience in many years. And just as I began to get carried away, the haunting echo of a distant memory returned. In the cath lab, the doctors and nurses wear lead suits to prevent radiation injury from the X-ray machines used to image the heart. The weight of this lead leaves a third of users with lower spine disease. At the end of my day, my heart would be full, but my back would throb. I tried to talk it down, to decatastrophize. But in the end, no matter how much I loved interventional cardiology, I knew that I was no match for the overwhelming force of chronic pain coming for me.[24]

We may work for basic sustenance, for food and shelter and health insurance, but we also work for meaning. Work provides us a way to be recognized, by our coworkers but also by society at large. For the fortunate, it can allow them to actualize their full potential, to be all they can be. For a diminishing few, it can even be transcendent, immortalizing.

Nothing can interrupt every step in this ascendant journey that work can offer like chronic pain.[25]

If someone who gets injured returns to work with chronic pain, the place they come back to is nothing like the one they left. Many people avoid showing any discomfort at work, worried that they may be seen as friable. Many worry about being perceived as work shy and creating slack for their coworkers to pick up as they are forced to cut back. Yet, if they minimize their symptoms, they may be asked to do tasks that could worsen their discomfort or may not get the flexibility they need. And no matter what they do, many face hostility or suspicion from their managers.

Returning to work may not even be possible for many with chronic pain. Musculoskeletal problems that render a person unable to work are skyrocketing. In 1960, disability was awarded to 17,124 people with musculoskeletal injuries, accounting for 8 percent of all workers given disability. Yet in 2019, even as so much manual labor has been replaced by less physically demanding work, 255,926 people were awarded disability for musculoskeletal injuries, accounting for 38 percent of the total number of workers awarded disability that year.[26]

Those unable to work because of pain are instantly otherized. They lose their connection to their own former selves, to the value their labor represented, and to the community their employment provided. No longer are they members of the universe they used to be part of. Many of them face tough choices: Should they go back to their previous job but risk not being up to par or should they look for a new type of job that may accommodate their physical limitations, even if they have to go down a few rungs for it?

Many people with chronic pain just stop looking, entering the increasingly large pool of people who are officially applying for disability.

Even as medicine has advanced, even as work has moved away from labor-based jobs to knowledge-based ones, and even as employers are required to offer considerable supports for people with disabilities, the number of people legally classified as disabled has seen a meteoric rise around the world. In the United States alone, about ten million people receive a disability check from the government, up from less than two million in 1970, at a cost of more than $250 billion annually. And although nominally a program for helping people get back to work, disability is almost always a one-way street: less than 1 percent of people on disability ever go back to work.[27]

Disability is not a specific diagnosis. Some people can work after heart attacks; others cannot. Some people work with pain while others either cannot or will not. Given that there is no test that you can run to show how much discomfort someone is in, no one really knows what makes chronic pain turn from debility in one to disability in another. This is particularly true of back pain, the most common reason people seek disability benefits. The degenerative spinal changes that were noted on the MRI I had performed on my back are extremely common in people with no symptoms of back pain whatsoever. Protrusion of a disc in the spinal column, the abnormality I had, sounds petrifying, but a third of twenty-year-olds have this finding noted incidentally on an imaging study of their back, even though they feel perfectly fine.[28]

Disability is a designation steeped in a dense economic, legal, and political milieu, and patients with chronic pain face considerable skepticism with regard to the legitimacy of their incapacity. Rand Paul, a Republican senator from Kentucky who also happens to be a physician, once told an audience in New Hampshire, "The thing is, all of these [disability] programs, there's always somebody who's deserving. [But] everybody in this room knows somebody gaming the system...[E]verybody over 40 has a little back pain."[29]

While acute pain might be one of our greatest adaptations, an essential element of our survival and growth, chronic pain can wreck the person it afflicts. Chronic pain can render people at odds with their social network, their work, the healthcare system—even with themselves.

When people with cancer are in the hospital, their rooms teem with flowers, friends, and family. Their stories are remembered and retold. Months of the year are dedicated to overcoming the emperor of all maladies. All of this recognition has generated hundreds of billions of dollars in research funding from governments, philanthropists, and the biomedical industry, which has fundamentally altered the course of cancer in our lives.

In contrast, chronic pain is the modern-day leprosy.

Chronic pain can sow doubt in patients about their own sense of reality. Their catastrophizing can often turn caregivers from advocates into instigators. Far from having work to fall back on, people in chronic pain find their work collapsing on top of them. And their lack of conformity to the rules of medicine can turn the healthcare system into an agent of persecution rather than therapy.

Despite its ubiquity, chronic pain remains deeply misunderstood. This misunderstanding exponentially magnifies the impact of the illness. We now know that chronic pain is not simply a prolonged form of acute pain and that it might well be mediated by an entirely distinct process. This emerging elemental truth about chronic pain might allow us to more deeply penetrate an infirmity that is both prehistorically ancient and pressingly modern.

4.

RAGE INSIDE THE MACHINE

The Fundamental Nature of Chronic Pain

Pain is as diverse as man. One
   suffers as one can.

            —Victor Hugo

CLIFFORD WOOLF'S OFFICE IS IN a tall, sleek research building in Boston's Longwood Medical Area. Just 213 square acres, Longwood is the mecca of modern medicine. It is home to Harvard Medical School and five of its affiliated academic medical centers, where almost three million patients are seen every year. More than a hundred thousand commuters descend on it every day. The research enterprise rakes in more than $1 billion in funding from the National Institutes of Health annually. The various labs provide a home to an estimated four hundred thousand zebrafish alone.[1]

While the labs in Longwood and in research facilities around the world have made tremendous progress in understanding conditions like heart disease and cancer, chronic pain has not traditionally been the focus of cutting-edge scientific investigation—until recently. Scientists around the world are now using novel techniques to uncover the fundamental nature of chronic pain. And their findings have shattered almost everything we know about this vexation and how we continue to treat it.

Woolf's corner office has a majestic view of this sprawling campus, which would have been even more impressive were it not dominated by ominous nimbus clouds the day of my visit. By the glass window sit several trinkets, many of which appear to be Southeast Asian in origin. One of them is a dragon appearing to float in the sky behind it.

Woolf grew up and went to medical school in South Africa. At that time, patients would often be left writhing in pain even after surgery. When an intervention was performed to provide respite, little effort was made to understand it. Woolf once asked a surgeon why he was using electrodes to perform a technique called transcutaneous electrical nerve stimulation to relieve a patient's pain. The surgeon replied, "Don't know, don't care, doesn't matter." Woolf saw a great gulf between the burgeoning vibrancy of the science of pain in research labs and how outmoded its clinical practice was in hospitals.

To close that gap, Woolf eventually made his way to the lab of Patrick Wall in London. Wall was the most famous pain researcher in the world and, along with his colleague Ron Melzack, wrote what remains one of the most influential papers in the short history of pain research.

Wall and Melzack's theory, first sketched out on a cocktail napkin in a Boston-area bar when both were at the Massachusetts Institute of Technology, outlined how people have come to think of both acute and chronic pain. It purported to explain why a soldier with a life-threatening wound did not feel pain or why rubbing a bruise helps to soothe it. Wall and Melzack theorized that pain signals were heavily modulated by gates in the nervous system. When a gate was closed, it halted all flow of pain; when it was open, signals could course through unabated. How wide open or closed a gate was depended on the factors twisting it one way or another: if someone was depressed, the gate would be ajar a lot wider than when, say, someone was approaching the end of a bike ride when the need to power through was principal.

The gate theory was met with initial resistance after it was published in *Science* in 1965, but it would ultimately help transform the study of pain from a soft to a hard science and is still widely taught. The gate theory is far from being considered the final word on the workings of pain, even by those who worked closely with its devisers. "Even though he was my great mentor, it's time to call a spade a spade," Woolf told me wistfully. "In those days you didn't need hard facts to get into *Science*. All you needed was an interesting idea."[2]

Still, Wall's evocative metaphor acknowledging the complex phenomenon as more than simply a bell-ringing alarm system would revolutionize the way we think about pain. When Wall died in 2001, from prostate cancer, Woolf—once his student and now one of the leading pain researchers in the world—wrote his obituary in *Nature*. "Some of the most successful contemporary biomedical scientists are really like chief executive officers of large multinational corporations, more involved in managing and delegating than in experimenting or thinking," he wrote. "Patrick David Wall, who died on 8 August aged 76, was the antithesis of this kind of scientist."[3]

When Wall was eight years old in England, a teacher told him that cotton grew in Lancashire. Yet, when he asked his parents, they said that while the county was famous for spinning and weaving, no cotton was grown there. In an autobiography, he described this as a watershed moment. "I was shattered by the revelation that some grown-ups in authority did not know what they were talking about, and I settled into a lifetime of doubting authoritarian pronouncements." Wall went to Oxford, where he was drawn to neuroscience. As an undergraduate student, he invented a rotating steel knife that could cut brain tracts with close precision. The results were published in *Nature*.[4]

A devout socialist, Wall was always drawn to fields of "social relevance," so it's no wonder he turned to pain. As a medical student he felt that patients' pain was greatly misunderstood and that "the explanations given to them and to me by my teachers were overt rubbish," he wrote. "The fantasy explanations often depended on mechanical disorders for which there was no evidence, such as trapped nerves, extra ribs, strained muscles, or floating kidneys. If those failed to convince even the doctors, there was a leap to using as an explanation the supposed inadequate personalities of the patients: neurosis, hypochondria, hysteria, and malingering."[5]

The scientific environment Wall graduated into was primitive. The prevailing view of pain was that specific nerve fibers carried nociceptive signals to a specific part in the brain, which was the pain center. Essentially, if a part of the body was hurt, it picked up the telephone and called up the pain center. While we know today that there is no "pain center," the idea was quite attractive at a time when people were likely to think of the brain as comprising different parts responsible for different activities. While this is true of senses such as vision, we know now it is not true for more complex phenomena like pain and consciousness.

As Wall entered the field, the established view of pain posited a linear 1:1 relationship between an injurious signal and how it was perceived: the more vigorously a nerve fired off nociceptive signals, the more pain the person felt. Wall was not a fan of this approach not just because of evidence to the contrary that he and his team found in the lab but because of personal experience he had in the world. During his trips to Hebrew University in Israel, where he had set up a lab, Wall saw many soldiers who had been mutilated in the Yom Kippur War. Instead of being in agony, many were completely pain-free. There was nociception but no pain. They were 1:0.

Wall also saw the ratio reverse to 0:1 in patients who appeared to feel pain without obvious active injury. This included patients who had strokes and later developed a condition called thalamic pain, which sent sharp, burning, and stabbing pain signals from the brain down the spinal cord and into the periphery in the absence of any noxious insult coming from below. These patients felt pain without any nociceptive trigger.

Wall and Melzack's theory proposed that a sentry manning the tollbooth determined the flow of signals up and down the spinal cord, invoking the metaphor of the gate. When the gate opened, hurt would flow; when it closed, hurt would halt.

This is why rubbing a bruise was soothing: the larger touch fiber signals from the rubbing overwhelmed the smaller pain fiber signals at the gate. The flow of pain was largely a matter of the balance between large- and small-fiber activity at the gate. But the gate could also be opened or closed by signals descending from the brain. This type of modulation could distract a badly wounded animal, for example, so that it could keep running to evade its predator. Emotions like joy or sorrow, euphoria or anxiety could also turn the gate one way or another.

The gate theory helped the theoretical science of pain reach its clinical reality. Pain was, the researchers wrote in their paper, "a linguistic label for a rich variety of experiences and responses."[6]

For all the influence Wall and Melzack's paper has had on pain research, the reason for its success was not the science it was based on but the graphic metaphor it conjured. "A fortunate aspect of our publication in 1965 is the use of the phrase 'gate control,'" they wrote in 1982. "It evokes an image that is readily understood even by those who do not grasp the complex physiological mechanisms on which the theory is based."[7]

"The physiology [that gate theory] was based on was so crude that all its predictions are wrong," Woolf told me. "It's not a balance of activity. It's not that you only get pain when the gate is open. You can modulate it normally and pathologically. But that's the fundamental flaw, and they [Wall and Melzack] were the climax of the pattern theory, but the pattern theory is dead." (The pattern theory posited that there are no unique receptors for elements such as heat, cold, and touch, which we know is not true.)

Wall's more enduring legacy was another line of work that he embarked on. "I was drilled in the classical view that the working mechanism of sensory systems was laid down in an immutable fashion during development and that no substantial functional changes could occur in the adult," he wrote. In other words, the nervous system that we were born with was thought to be exactly the one we died with, as static as our fingerprints. Wall made one of the first forays into the brave new world of neuroplasticity, setting the stage for his protégé Clifford Woolf to discover just what keeps the spindle of chronic pain spinning and spinning and spinning.[8]

Woolf presents as a cross between a Silicon Valley futurist and a Tibetan monk. He has a clean-shaven head with a faint goatee. When he moves his thin frame, he does so almost weightlessly. He speaks softly and dresses sharply yet casually, wearing a black T-shirt and jeans. He often looks into the distance as he speaks.

Woolf doesn't see patients, but he spends a lot of time thinking about how clinicians manage chronic pain. "We basically treat chronic pain the way we used to treat tuberculosis before we discovered what caused it."

Tuberculosis has been known by many names—consumption, phthisis, white plague, the king's evil. It shows up in Egyptian papyri, in the Old Testament, and in the ancient Hindu Vedas. In the nineteenth century, every fourth Londoner died of it. Its inevitability made it sought after among the European elite, many of whom would make their skin look pale to mirror its effects. While tuberculosis was determined in 1882 to be an infectious disease caused by mycobacteria, the first effective treatment for it, the antibiotic streptomycin, wasn't discovered until 1944.[9]

In the sixty-two years between 1882 and 1944, as the toll of tuberculosis kept rising, people desperately reached for whatever they could to stave off its scythe. Lungs were cut out. Nerves were crushed. Sanatoriums were built in mountainous resorts around the world to house the consumed with the hope that fresh air and healthy diets would whisk the contagion away.

This, to Woolf, is precisely where we are when it comes to understanding and treating chronic pain today. "If you go into the emergency room, they have standard pictures of faces crying to help patients communicate what they are experiencing," he said. "I can't think of a cruder measure of assessing pain."

Even sophisticated physicians for the most part still differentiate pain simply based on where it occurs—in the back, in the neck, in the head, in the joints—rather than the mechanism causing it. When the pain arises from the nervous system itself, it is known as neuropathic pain. Our peripheral nervous system is the sentinel of our body, always on the lookout, sending messages back to the center as it detects disturbances. When it comes to neuropathic pain, the messengers become the menace.

Nociceptive pain—the kind of pain we feel when we prick our finger or receive a punch to the gut—occurs when lesions are detected by the sensory system, which is made up of nerves, the spinal cord, and the brain. Neuropathic pain, on the other hand, occurs most often because of damage to the sensory system itself. Instead of firing in response to an external or internal threat, the system fires randomly. Both large-bore A fibers, which transmit pain quickly and precisely, and the smaller, finer C fibers, which transmit pain slowly and more diffusely, can be affected.

Neuropathic pain feels different than the usual type of pain. It is described as burning or tingling; even light touch can hurt as the body becomes exquisitely tender. Many people with nerve damage, even though they feel pain, paradoxically also feel numbness. How can numbness and pain coexist? If a nerve fiber's connection with the spine gets severed, the brain stops receiving sensory information from the affected area. However, the stumps of the remaining pain fibers in the now anesthetized area become fidgety and irritable, firing off erroneous SOS signals back up to the brain.

The most common cause of neuropathic pain is peripheral neuropathy, which often affects the hands and feet symmetrically—the so-called glove and stocking distribution. This occurs because such neuropathy most commonly affects the longest nerve fibers, which then begin to die back, leading to loss of all sensations and, less commonly, causing weakness as well. This is the classic pattern seen in the 50 percent of people with diabetes who experience neuropathy. In others, neuropathic pain can be very regional. Pressure on a nerve that innervates the face can cause trigeminal neuralgia, causing sharp, intermittent, and sudden bouts of pain in the face. Another condition that causes neuropathic pain is shingles, which affected me and the patient I had mentioned earlier. Shingles occurs when dormant herpes virus reactivates in the peripheral nerves, resulting in pain in the region of the affected nerve long after the virus has been successfully treated. After surgery, people can develop neuropathy in the parts of the body that underwent the operation.[10]

Another common cause of peripheral neuropathic pain is when a nerve gets pinched as it leaves the spine. These "radiculopathies" occur often in the lower back or in the neck. The numbness, tingling, and shooting pains are felt in the regions supplied by the nerves involved—the arm when the squeeze is in the neck, the leg when it is in the lower spine, a condition also called sciatica.

There are several other reasons why people might develop peripheral neuropathic pain, many of which can be very difficult to diagnose. Lesions in the nerves can alter the ion channels in the nerve membranes—these channels, through the exchange of electrolytes, generate the currents that spark neuronal signals. In rare circumstances, people can be born with mutations in these channels. In one such condition, paroxysmal extreme pain disorder, people have episodic burning in their eyes, their jaws, and their behinds. Infants born with these mutations can cry inconsolably when they are eating or pooping and often go undiagnosed. Even the thought of food can at times bring on the agony.

Drew Leder, the physician who eventually became an anthropologist and philosopher, had a long history of back pain. "I had a herniated disc, and I avoided surgery at all costs. I went to a chiropractor and had a number of epidurals," he told me. Leder's attempts to stave off the scalpel failed spectacularly. "I collapsed on the hotel floor. I couldn't even reach the telephone to call for help. The disc had broken off and was traveling down my spinal cord. When a neurosurgeon took a look at it, he took me into surgery immediately."

For the most part, Leder had a good response, but he went on to develop a mysterious form of chronic pain. "It was just above my ankle on my left leg. It was a very stabbing pain that sometimes would be worse, sometimes less. I am a big walker, often walking two hours a day. The pain would only allow me to walk for ten or fifteen minutes," he said. "At times it became overwhelming. I would have to lie down, elevate my leg and use forms of mental imagery to distract myself."

The pain lasted for years as Leder shuttled between different physicians. "The worst interactions were when specialists threw their hands up in the air and said there was nothing they could do for me. It left me feeling very unheard and in despair."

Despite the disappointments, Leder clung to hope. But that wasn't necessarily a good thing.

"Hope can be poisonous," he said. "Giving up hope that things will be otherwise can allow people to come into a phase of acceptance and can give the individual psychological and physical relief. They begin to work within the parameters of what is possible for them as opposed to trying to defeat the illness."

The insatiable power of hope, though, is pervasive, even among the prisoners Leder studied. "Pretty much to a man, people with life sentences hope they will one day be free men. Many believe that if they give up all hope, they will become vegetables."

Yet this same hope for a diagnosis drove Leder from doctor to doctor and ultimately brought him to be diagnosed with an idiopathic neuropathy involving the saphenous nerve, a long nerve that collects sensory information from most of the leg. The term "idiopathic" is geek speak for "we have no idea what causes this." For anyone seeking to wrap a medical condition into the fabric of their narrative, to provide their suffering with an origin story, and then to move forward and potentially pursue treatment, "idiopathic" can be a defeating term. Given how little we know about pain, especially chronic pain, it is not uncommon for patients to wear the dreaded idiopathic tag around their necks.

What we know about idiopathic neuropathy is that it occurs mostly in people in their sixties and can cause both numbness and pain. While some can find relief with medications used to alleviate neuropathic pain, none of those appeared to be working for Leder until a plastic surgeon proposed a dramatic strategy: a nerve transplant.

An option for some patients who don't respond to medication is to dissect the affected nerve. If there is no nerve, there is no way for those abnormal pain signals to be generated. However, the surgery risks deadening the region the nerve serves, creating more problems than it solves. To overcome this, some surgeons can replace the diseased nerve with a healthy nerve from another part of the person's body. If no other nerve can be used, as in Leder's case, surgeons can transplant a nerve from someone else or from a cadaver. Desperate for a solution, Leder tried the nerve transplantation.

The surgery didn't eliminate his pain, but it did force him to reconsider his relationship with it. Having now attempted the most aggressive form of treatment he could pursue, with hope for a cure no longer leading him in circles, Leder knew he had to learn to live with the pain. Instead of hoping to be completely rid of it, Leder changed how he spoke to it.

"This part of my body is trying to tell me something," he said to himself. "It's not my enemy. It's like we are two friends going along in life together." Instead of shunning its presence, accepting pain as part of his life actually helped him more than any other intervention had.

Lesions in the nerves are not the only conditions that cause neuropathic pain. Some disorders of the central nervous system (i.e., the brain or spinal cord) can as well. This includes illnesses like Parkinson's disease or multiple sclerosis, as well as the aftereffects of stroke. Central neuropathic pain can occur because the brain has an important role in not just generating the pain experience but also inhibiting it.

Inhibitory buffers in the brain and spinal cord allow us to ignore information deemed extraneous. For a weight lifter, it may involve ignoring the acid building in their arms and the clanging shocks in their knees. For a writer, it might involve ignoring the soreness in their back, the tense grip around their neck and shoulders, and the dull pull of gravity on their eyelids as they type away into the wee hours. Even as I write this and consciously scan my body sitting on this chair, I become instantly aware of multiple pressure points in my back and legs, as they prop me up.

Central neuropathic pain marks the complete reversal of how we have experienced pain throughout our history. Pain has always been thought of as a bottom-up phenomenon, with tissue damage detected at the cellular level and transmitted up the chain of command all the way to our brains. Central neuropathic pain, though, is a top-down phenomenon: the brain re-creates pain without any new signals asking it to do so. It's almost as if Hippocrates, the Greek physician whose oath is recited to this day the world over by medical students, knew something long before anyone else when he wrote, "Men ought to know that from the brain, and from the brain only, arise our pleasures, joys, laughter and jests, as well as our sorrows, pains, griefs and tears."[11]

Neuropathic pain challenges the simplistic notion that pain is only a product of external threats. Yet no form of neuropathic pain shakes up our understanding of chronic pain more than its most enigmatic archetype: phantom limb pain.

Phantom limb pain came to prominence after "The Case of George Dedlow" was anonymously published in the *Atlantic Monthly* in 1882. George Dedlow lost all his limbs, one by one, during the American Civil War and was left "dead except to pain." Donations poured into the Philadelphia facility known colloquially as "Stump Hospital," where he along with countless other soldiers with amputations were being cared for. Some wrote him letters; others even visited the hospital hoping to meet him in person. He became the face of a war that had cost 750,000 lives, leaving innumerable amputee veterans behind as its most visible legacy.

"The Case of George Dedlow" remains one of the great descriptions of human resilience in the face of suffering ever written. Yet, as the people who showed up to Stump Hospital found out, no person by the name of George Dedlow ever existed.

What made the American Civil War particularly gruesome was the minié ball, a bullet widely used in front-loaded muskets. Unlike the smooth, round lead balls that preceded minié balls, these conical, hollow bullets were accurate at long ranges. They also flattened like pancakes as soon as they hit a human body, plowing large, vacuous tracks through their victims, unlike round bullets, which often went straight through. They caused wanton destruction of bone, blood vessel, fascia, and ligament, often leaving amputation as the only recourse. The surgeries that amputees underwent were often worse than their injuries. Surgeons frequently operated in "old blood-stained coats...with undisinfected hands...[using] marine sponges which had been used in prior pus cases and only washed in tap water." Unsurprisingly, these amputations led to death in a quarter of soldiers and rip-roaring gangrenous infection among survivors.[12]

The minié ball left its horrific mark on countless men, the families they returned broken to, and the nation left to pick up the pieces. Deeply tainted too were the army surgeons, whose humanity was drained by their proximity to the violence and who became known for "inefficiency, gross carelessness, heartlessness, and dissipation."[13]

One man deeply affected by the gory legacy of the minié ball was Silas Weir Mitchell, the third of nine children, who spent his entire life in Philadelphia. His father was an exacting physician who was also fond of literature, and Mitchell followed in his exact footsteps, eventually taking over his medical practice while also writing poetry and prose. Notably, he spent two years under the tutelage of Claude Bernard, a French experimental physiologist and one of the enduring pioneers of the scientific method. Here an interest in neurology was fermented.[14]

When the Civil War broke out, he served the North as a contract surgeon. He used his influence with the surgeon-general to open the Philadelphia hospital dedicated to soldiers with nerve injuries, paralysis, and seizures that came to be known as "Stump Hospital." A year after the Civil War ended, Mitchell wrote the fictional story of George Dedlow. According to later recollection, unbeknownst to him, the draft reached the father of a friend, who submitted it to the magazine without his knowledge.

Told in the first person, George Dedlow's story starts off with the disclaimer that the account had been turned away "by every medical journal" to which he had submitted it. While he was not a physician, the fictional Dedlow had some medical background and was stationed at a medical post in the war. Over the course of his service, he was injured multiple times, which led to all four of his extremities being amputated one by one. The amputations were initially welcomed. "The pain [of the operation] was severe but it was insignificant compared to any other minute of the past six weeks." When he saw one of his severed arms on the hospital floor, Dedlow said, "There is the pain, and here am I." The ability to shed parts that knew only pain gave him hope.

And yet, as his organic pain died, its ghost came to life. Once after surgery, he was awakened in the middle of the night by cramps in his legs. He was too weak to help himself, so he called out to an attendant.

"Just rub my left calf," said I, "if you please."
"Calf?" said he, "you ain't none pardner. It's took off."
"I know better," said I, "I have pain in both legs."
"Wall, I never," said he. "You ain't got nary leg."

The attendant threw the covers off, revealing to Dedlow's horror that both had been sawed off, even as the hurt from them was twisting him in knots. He was eventually taken to Stump Hospital, where he was now one of many hundred limbless men. There he "found that the great mass of men who had undergone amputations, for many months felt the usual consciousness that they still had the lost limb. It itched or pained, or was cramped, but never felt hot or cold. If they had painful sensations referred to it, the conviction of its existence continued unaltered for long periods."

Mitchell did not reveal himself as the author of the story until a few decades later. Meanwhile, he published academic volumes providing insights into this shadowy state. One of Mitchell's patients had lost his right hand at the Battle of Gettysburg. Even four decades later, the memory of his hand returned, but only in his dreams. "I write often in my dreams," his patient said. "I attempt to use the tendons which would hold and guide the pen." Yet his dreams would quickly become nightmares as his hand would refuse to comply; instead his tendons would cramp, "wakening [him] up from the most profound sleep because of pain."[15]

While he gave it the name, Mitchell was far from the first person to describe phantom limb pain. In Western literature, that distinction goes to Ambroise Paré, whose introduction to phantom pains was also facilitated by gunshot wounds he treated in his work as a French military surgeon. At a time when surgeons were widely considered uncivilized and had a stature similar to butchers and barbers, he established the profession as a scientific vocation and, despite being born to low station, served as the chief military surgeon for four French monarchs.[16]

As if amputation without anesthesia were not bad enough, in those days after a limb was amputated, the stump was burnt with a searing red-hot iron and then doused in boiling elder oil. Few actually survived this brutality. Paré developed a new ointment that included the antiseptic turpentine and resurrected the ancient Roman technique of tying off severed blood vessels and closing wounds with sutures rather than burning them. Not only did the patients do better and suffer less infection, but they hurt a lot less too. Yet, because so many more survived amputations, he also began to notice that phantom limb pain was common.[17]

In one of his many textbooks, Paré wrote that "this false and deceitful sense appears after the amputation of the member; for a long while after [patients] will complain of the part which is cut away." He was cognizant many would find this account fantastical. "Verily it is a thing wondrous strange and prodigious, and which will scarce be credited, unless by such as have seen with their eyes, and heard with their ears the patients, who have many months after the cutting away of the leg, grievously complained that they yet felt exceeding great pain of that leg so cut off."[18]

Phantom pain became a well-known phenomenon in France and was described in great depth by a recurring figure in the annals of pain, René Descartes:

> I once knew a girl who had a serious wound in her hands and had her whole arm amputated because of creeping gangrene. Whenever the surgeon approached her they blindfolded her eyes so that she would be more tractable, and the place where her arm had been was so covered with bandages that for some weeks she did not know that she had lost it. Meanwhile she complained of feeling various pains in her fingers, wrist and forearm; and this was obviously due to the condition of the nerves in her arm which formerly led from her brain to those parts of her body. This would certainly not have happened if the feelings or, as she says, sensation of pain occurred outside the brain.[19]

Phantom pain became the linchpin of what would come to be known as the idea of Cartesian dualism, a worldview that drew a hard line between animals, who could experience sensations but not perceptions, and humans, who experienced both. Because phantom pain could be experienced without the presence of a leg or arm to originate it, Descartes reasoned that pain, a faculty of humans alone, lived only in the human brain.

Although phantom limb pain appeared to be a well-known

condition and became particularly prominent after Silas Weir Mitchell's short story and his subsequent academic work, it took decades
before the medical establishment recognized it as a legitimate medical diagnosis. In a case series of more than a hundred patients
seen at the Mayo Clinic published in 1941, physicians Allan Bailey
and Frederick Moersch described the spectrum of phantom pain
in amputees. The pain was described as of a "burning, aching or
cramping type" with a "crushing, twisting, grinding, tingling, tearing
or drawing quality." For some it came on as soon as they lost their appendage; for others it first emerged decades after. Some said the pain
was aggravated by changes in the weather but relieved by drinking
alcohol.[20]

The irony of phantom pain is that those who have it are far more
aware of their limb after it is gone than when it was still attached to
them. The absent limb is made so immediately present by the pain
that it can become unbearable, driving some even to suicide. The
paper by Bailey and Moersch provided the final push that helped
legitimize phantom pain as a real diagnosis; it was inducted into the
Index Medicus, the official list of medical terms, in 1954.

Of all the different forms of chronic pain, the sensory ghost of
phantom pain most starkly brings the core nature of pain into dispute:
Is the diagnosis of phantom pain a psychological one, like major
depression, or a physical one, like diabetes?

Bailey and Moersch were quite certain which side they were on:
phantom pain was psychic in nature and was a form of "obsessive
neurosis." Many doctors after them, however, would come to different
conclusions.[21]

George Dedlow, who felt like "a useless torso, more like some
strange larval creature than anything of human shape," believed that
he had lost his identity. "I found to my horror that at times I was
less conscious of myself, of my own existence, than used to be the
case...This set me to thinking how much a man might lose and
yet live."

One day he skeptically visited a medium with a spiritually inclined fellow patient. The alphabet cards he touched caused the medium to say, "United States Army Museum, Nos. 3486, 3487."

"Good gracious!" shouted Dedlow, "They are my legs! My legs!"

"Suddenly," he recalled, "I felt a strange return of my self-consciousness. I was re-individualized, so to speak." He rose up and staggered across the room "on limbs invisible to them or me." His brief reassembly with his legs, which had apparently been stored away in alcohol-filled vats as specimens in a museum, was as fleeting as it was euphoric as he "fainted and rolled over senseless." Eventually, he was discharged from the hospital to his home in Indiana, "not a happy fraction of a man...eager for the day when I shall rejoin the lost members of my corporeal family in another and a happier world."

It may seem astounding that so many would take as truth a story that ends so outlandishly, with legs restored momentarily in a séance. The widespread belief in the veracity of Dedlow's fictional account perhaps reflected the suffering of those who found a friend in a human torso.

War and disease have continued to rage on, with bullets and bombs continuing to rain, ripping people away from the components that make them whole. The voids that these losses leave behind, the memory of legs and arms that only remember pain, persist. What has changed, however, is how well we understand phantom pain and what it teaches us about the pains we have in the parts of our bodies that are present.

No one really knows why people have phantom limb pain. There are "hundreds of theories in the literature and few or none are capable of being tested rigorously," according to researchers in a recent review of the phenomenon. We do know that people with amputations can have disruptions starting from the site of their stump, through the nerves that connect it to the spinal cord, and even in the innermost recesses of their cerebral cortex.[22]

When a limb and its associated bones, muscles, tendons, and blood vessels are scythed off, so too are the nerve fibers coursing through them. These damaged neurons can start emitting erratic signals that travel up the spinal cord and cause pain. The tissue around the injured nerves can thicken, resulting in the formation of neuromas. It would stand to reason that anesthetizing these damaged nerves and neuromas would put an end to the phantom pain. Yet research suggests doing so makes no difference. Furthermore, while it may take time to develop a neuroma after an amputation, phantom limb pain can begin right after the operation.[23]

For that reason, phantom limb pain is increasingly seen as a top-down phenomenon.

Imagine the pain network in the brain as a workplace gossip mill. In a relatively benign workplace, the gossip mill is fed with actual tidbits of information, which would be analogous to noxious stimuli whispered from one ear to the next. Maybe someone saw Jack whispering to Jill by the coffee machine and presumed they were having an affair. Yet, in a toxic workplace, the gossip mill doesn't just turn innocuous information into salacious scandal; in many cases it does not even need any input to get started. Rather, it is so diseased that it constantly churns out rumors and misinformation without any antecedent basis. Jack and Jill don't even need to be in the same country together to have their reputations tarnished.

The connection between the mind and the body has long been thought to be airtight. This is precisely what has made phantom limb pain an object of so much fascination: How can a part of the body not there still be yours? And yet you don't have to lose an appendage to know that the relationship between body and mind doesn't always run smoothly. The human brain can be as toxic and malignant as any workplace gets.

Case in point is the rubber hand illusion, in which a person's own hand is concealed—say, behind a piece of cardboard—and a rubber hand is placed right next to it, which the person can see. Now both the visible rubber hand and the invisible actual hand are touched simultaneously—say, with the stroke of a brush. In time the rubber hand comes to seem like it is the person's own hand as his or her consciousness tries to occupy it. When a hammer falls on the rubber hand, people actually feel pain, and many shriek and jump in shock.

The experiment shows that our nervous system is malleable, rapidly taking the shape of whatever vessel it needs to resemble.

As you read this book, how do you know where your legs are? Surely you weren't consciously thinking of them before I forced you to attend to them. The awareness of where our bodies lie in three-dimensional space is called proprioception. Even when we lose our limbs permanently, a proprioceptive engram denoting a virtual three-dimensional map of the lost limb can still remain in our brains. At times, amputees can feel their missing limbs stuck in twisted positions, often in the last position they were in before being severed.[24]

People who have pain before an amputation are far more likely to have pain afterward than those who don't. Furthermore, people who are born without limbs and therefore have no memory of an injury are far less likely to have phantom pain from a limb they never had, while those who lose one to an amputation are much more likely to experience discomfort.[25]

Dreams can connect us to our inner reservoirs of memory in both erratic and lucid ways. They offer us not just an alternate series of experiences but a different way of looking at ourselves. Most people, when dreaming, see themselves floating over a scene, observing it, which makes dreams a useful tool for studying phantom limb pain.

How do we see ourselves in our dreams? Researchers argue that we all are born with an innate map of what a normal human body comprises. They point to people born with paraplegia, who therefore have never had control of their legs and have no prior memory of what it must be like to walk on them. And yet, even those born with paraplegia see themselves walking in their dreams.[26] But what do those who have lost their limbs see?

In a nationwide survey of German amputees called PHANTOM-MIND, 25 percent of respondents always saw themselves as physically intact in their dreams, while only 3 percent always saw themselves in the impaired frame they inhabited in the present. The others either didn't remember what they looked like or alternated between impaired and intact. Though they were few, the people whose dreams featured their amputated bodies actually were more likely to have phantom pain.[27]

Phantom pain occurs at the intersection of dreams, memories, and reality. In their dreams, amputees are intact. In their dreams and memories, amputees can hop on one leg or shoot a basketball with one arm, not because they have to but because they want to. Amputation, one of the greatest breaches of one's bodily integrity, creates a cataclysmic severance between the innate selves of our dreams and memories and our real selves. And it seems like the people who accept their bodily disruption, those who see themselves as cleaved in their dreams, are even more likely to suffer from their limb's ghost.

Phantom sensations don't just haunt those with missing limbs. Brian woke up in the middle of a night with what he felt was an erection. While he initially thought he was dreaming, when he pushed his hips forward, the bulge in his boxers flattened out. It had been a year since Brian had his penis surgically removed because of a cancerous growth; ever since, Brian had had a phantom penis. For those with phantom penises, the experience is a morbid mixture of pleasure and pain. The phantom penis responds to sexual arousal. But the inability to act on it and the memory of the loss it rekindles can be discombobulating.[28]

The extremely provocative question that phantom pain has long suggested is whether all chronic pain is phantom. While phantom limb pain is marked by a visible loss, causing researchers to use mirrors and virtual reality to provide an illusion of restoration (to little avail), the loss in most patients with chronic pain is far less visible.

Often those with chronic pain don't even have a diagnosis, let alone an evident injury. Yet phantom pain and chronic pain might have more in common than meets the eye.

Much of science is driven by hypotheses. Once a hypothesis is in place, experiments are designed and performed to prove or disprove it.

Clifford Woolf spent long hours in Patrick Wall's lab at University College London. What he did there would conventionally be described as discovery science and callously called a fishing expedition. Patrick Wall's persona, like the lab that embodied it, was a perfect incubator for scientific investigation not bound by the rigidity of hypotheses.

Rigidity, though, was the core tenet of how the human nervous system was understood at that time: the prevalent belief was that the nervous system you were born with was the exact same one you died with; that the nervous system was fixed and incapable of change.

We know this not to be true today: the human brain is incredibly *plastic*. But how did we get here? "In contrast to hypothesis-driven science, discovery science is an exploration of the unknown," Woolf wrote in an article for the journal *Anesthesiology* in 2007. "There are no road maps from the National Institutes for Health, just narrow, twisting paths, many dead ends, and very occasionally, a totally unexpected byway."

Instead of focusing on the fuzziness of the pain experience, Woolf concentrated on one of its most tangible aspects: the withdrawal reflex. You grab a hot pan steeping chai, and you immediately jerk your hand back. Depending on how hot it is, you might even spill chai all over the stove and floor (and you will be reminded of your amateurism and lack of foresight by your spouse until the end of time).

Woolf was measuring pain reflexes in rats whose brains had been removed but whose spinal cords were intact. He noticed that the muscles' responses to pain seemed to be quite erratic. At times stimulation would result in a timid withdrawal; at other times the withdrawal was much brisker. Not only did the sensitivity vary, but the area of the body that the muscles responded to changed. At times a particular muscle would withdraw if a toe was pinched, but at other times it would withdraw if any part of the leg was pinched. For many months Woolf toiled in search of an answer until he realized that the time of day he was performing the experiments was changing. When stimulated toward the end of a long day during which they had been stimulated many, many times, the muscles would withdraw from the painful stimulus much more vigorously than when they were stimulated earlier in the day. Muscles also appeared to respond to painful pokes across a much wider area of the skin later in the day after they had been excessively stimulated.

Woolf could have brushed this off as an error in measurement or simply an artifact of his own long labor in the lab. Instead, he took this serendipitous finding seriously; it would become one of the most important discoveries in understanding the mechanisms of chronic pain. Woolf discovered that the repetitive stimulation that he was subjecting the spinal cord to was actually changing the way it transmitted pain: the more the nerves were stimulated, the more responsive they became. Rather than becoming exhausted or dulled, as one might assume, the more pain they transmitted, the more exquisitely sensitive the nerves became. Woolf called this phenomenon "central sensitization" and published the finding in 1983 in a rare single-author paper in *Nature*. Subsequent work he performed showed that not only does the function of the nerves change with chronic pain, but the very anatomy of the nerves and their connections undergoes a rearrangement.[29]

Woolf proved that pain begets more pain. While conventional thinking has always suggested that those who experience pain often become more used to it, the opposite happens: people with chronic pain are more sensitive to pain than those who don't have chronic pain. Whereas the conventional maxim claims that what doesn't kill you makes you stronger, it turns out that the more one hurts, the more sensitive one becomes.

And the longer chronic pain lasts, the more the human nervous system evolves. In fact, if chronic pain persists long enough, it changes not only our nerves and spinal cord but the wiring and connections in our brains. Before too long, chronic pain begins to look nothing like acute pain at all. When you first hurt your back, your throbbing rear sends SOS signals hurtling to the brain. Yet, in chronic pain, it is the brain sending alarm signals down to the back. Pretty much the only thing that might be common between these two is how they make someone feel.

How acute pain shape-shifts into chronic pain can help us understand how they are different.

One study performed at Northwestern University used fMRI scans to compare patients with acute and chronic pain, as well as acute pain patients who went on to develop chronic pain. They found that acute and chronic pain appeared to light up two separate, nonoverlapping circuits in the brain. The signature of acute pain in the brain, they found, was quite different from that of chronic pain. When they looked at those patients who transitioned from transient to persistent pain, they saw that change happening in front of their eyes: while acute pain was initially localized in parts of the brain largely involved with sensations, over time the circuits shifted to those primarily regulating emotions. Further research has confirmed this: a study that compiled the results from fifty-one different brain-imaging studies showed that the anterior insula, primarily responsible for assigning emotions, is the node that undergoes the most significant change in chronic pain.[30]

The more scientists have investigated chronic pain, the more distinct from acute pain it appears. The shape-shifting nature of chronic pain means that it doesn't fit neatly with anything else we have traditionally compared it to. It is certainly not a physical sensation, and yet it doesn't necessarily behave as an emotion either.

Perhaps chronic pain is something else. The young boy with the heart transplant, traumatized by being awake under anesthesia, experienced pain in a more visceral and violent manner than anyone else I have seen. But how much of what he felt was pain and how much was simply the progressively louder echo of traumatic helplessness?

We humans have developed elaborate mechanisms to bury difficult moments deep within the subconscious. The emotional wound of abuse or abandonment as a child, even when overtly forgotten, can manifest as a lifetime of anxiety and posttraumatic stress disorder. But what happens to physical injuries? What happens when we try to brush their recollection under the carpet, hoping they will remain buried forever?

What if chronic pain is neither a physical sensation nor an emotional state? What if chronic pain is something else altogether: a memory?

Before we had papyri to scribble hieroglyphs on in the sand dunes by the Nile, diaries to fill as we lay in bed about to fall asleep, or social media to record every moment of our waking lives, our only means to document our lives lay in our brains. The commitment to memory of euphoric victories and cataclysmic downfalls, of the sweet burst of a tangerine in our mouths and the acidic lashing of a hot pepper on our tongues, formed the very basis of how we came to live.

Yet human memory is not a tape recorder. It picks and chooses, subduing and amplifying what it wants to store. How we feel in the moment can affect how we remember. We may recollect being robbed of our cellphone more strongly than simply misplacing it. Bland food slips into oblivion, but the memory of tear-inducing spice is always retrievable.

Negative emotions can easily turn memories into deep wounds that might never fully heal. Merely remembering a traumatic burn can bring with it the smell of burning flesh and the blistering of the skin. And memory retrieval isn't a passive process—it can indelibly change the substance of the recollection. In fact, while retrieval of emotional memories can bring back extremely vivid images, these retrieved memories can be very fluid, malleable like wet clay.

How do we put together our catalog of sights and sounds, of sensations and emotions? Long-term memories are stored via the connections between the neurons in our brain. Given that neuronal signals are fleeting, long-term remembrance necessitates alterations in gene expression in the involved cells. The development of these strong neural networks in the brain is brought about by a process called long-term potentiation. At the heart of the commitment to memory through long-term potentiation is a protein called PKMzeta.

The gaps between neurons are called synapses, and they are the glue that holds our nervous system together. Consolidating memories requires strengthening the synaptic connections between specific neurons, which is exactly what PKMzeta does.[31]

The memorialization of past events by PKMzeta is not permanent. In fact, left untouched, our memories continue to decay, unless we access them again through retrieval. Our memories are like the torn pages of a journal being flung around by gale-force winds as we desperately flail to hold on to them. An infusion of PKMzeta, however, can suddenly solidify fast-evaporating memories in rats, scientists have shown.

PKMzeta does more, though, than just aid us in remembering the passwords to our online accounts. It is essential to the genesis of chronic pain.

PKMzeta primarily shows up in the hippocampus, the part of the brain devoted to memory and learning. Yet researchers have found a wider footprint when it comes to its relationship with pain. A landmark experiment published in the journal *Science* showed just how similar the mechanisms underlying chronic pain and memorialization are. Injury to peripheral nerves increased the levels of PKMzeta in the parts of the mouse brain receiving nociceptive signals from those nerves. These injuries led to the development of chronic neuropathic pain in the mice, manifest as a phenomenon called allodynia, when even nonpainful stimuli hurt. Yet, when researchers injected ZIP, a substance that inhibits PKMzeta, effectively blocking the ability of

PKMzeta to commit events to memory, the mice no longer developed behaviors characteristic of chronic pain.[32]

Therefore, as mice lost the ability to remember, they also lost the tendency for acute pain to be learned and recalled as chronic pain.

A subsequent study by a different group of scientists showed that PKMzeta also helps plant an engram of chronic pain in the spinal cord and that the injection of ZIP after an injury can reverse the sensitization that develops after acute pain, preventing it from becoming permanent pain.[33] Based on this work, we can say that in mice acute pain becomes chronic when it becomes memorialized, allowing it to be felt even when no further injury is present. But does this hold true for humans as well?

Vania Apkarian was an electrical engineer before getting a PhD in neuroscience. At the beginning of the 1990s, when fMRI research had just begun, he combined his training in engineering and biology to apply that technology to study pain. Apkarian's approach, however, received a lot of pushback. Traditional basic scientists are used to studying animals in very controlled settings, where they can make very precise modifications that would be impossible in human beings. Scientists who were studying humans were largely using healthy volunteers and exposing them to a transiently bothersome sensation. Apkarian decided to instead focus on patients with actual acute or chronic pain. This seemingly intuitive decision came with its own challenges. "It is much harder to publish a paper on patients rather than healthy subjects," he told me. "Patients don't necessarily do what we want them to do."

Apkarian is a professor at Northwestern University and, according to one British researcher, has "received enough grant money to support a medium sized country." More than any other person I know, Apkarian has laid the groundwork that is pushing people to rethink chronic pain. His decades of research have transformed him from a little-known outsider into a very well-known outsider in the science of pain. "The drug industry listens to me very intently, they take a lot of notes, and then they disappear and don't want to talk to me," he said. "I used to be a consultant to many drug companies. I would tell them they are not doing the right research, and they wouldn't invite me again."[34]

Pain doctors shun him as well. "My own anesthesia department doesn't want to hear from me," he said. When he does get a chance to speak to physicians, he recalls often being asked, "Do you mean what we are doing is not the right thing?"

Apkarian's work has upended how we study chronic pain, challenging, among other things, researchers' and clinicians' obsession with dissecting the origin story of persistent pain, its initial trigger. "The classic idea is that if the injury is bad enough, it will stay on," he told me. "The injury itself has no value."

Apkarian has shown that the intensity of the initial injury has little to no connection with whether it will die away or become a perpetual pestilence. It was also Apkarian's lab that performed the groundbreaking research finding that chronic back pain, unlike acute pain, largely involves the emotive parts of the brain. Pain, to Apkarian, is an emotion you can localize to a specific part of your body. His current research is an attempt to see if the real root of chronic pain is our remembrance of it: "Painful stimuli are the most effective means to develop memory."

The hippocampus is the part of the limbic system specific to learning and memory. Activation of specific parts of the hippocampus leads us to recall specific memories, such as the fear of an electric shock in a mouse. Damage to it can impair the ability to both recollect and to make new memories. Emotion and memory coexist in the hippocampus, which means our recollections aren't like camera recordings but are influenced by the feelings we have about them.

Pain leaves an indelible imprint on the hippocampus. While the memory of pain can often last a lifetime, the hurt we retrieve can often be quite different from that which was felt. For example, two famous experiments by Donald Redelmeier and Nobel Prize winner Daniel Kahneman showed that when thinking back about a painful procedure, people were most likely to remember the pain at its worst and during its final moments.

In one experiment, the researchers recorded patients' pain ratings in real time as they underwent two types of uncomfortable procedures. When asked to recall the experience afterward, the remembered intensity of discomfort correlated most strongly with how bad the pain was at its peak and how bad it was in the last three minutes of the procedure.[35]

In a subsequent trial, patients either underwent a standard colonoscopy or a colonoscopy at the conclusion of which the tip of the probe was left in place in the patient's rectum for a few extra minutes. While this additional step prolonged the test, it did result in the last few minutes being less painful. Not only was this longer test rated as less uncomfortable overall, but the patients who received the longer test were also more likely to return for a repeat colonoscopy. This feature of how we recall pain at its worst and how it ends is called the peak-end rule.[36]

We remember pain only at its worst, biasing our recall of agonizing encounters to their extremes, because the more negative emotion an event generates in our brains, the more solidly cemented it becomes in our mental museums. The distant echo of pain is also amplified by its witnesses. After children undergo surgery, their recollected discomfort is amplified more by their parents' catastrophizing than their own. Any additional attention that children receive during acutely painful episodes results not only in their experiencing greater pain in the moment but in the impact lingering long after, since children with more distorted memories are more likely to experience more pain and develop chronic pain in the future.[37]

You would think that those with persistent pain would be the best at remembering their pain, but research suggests the opposite: people with chronic pain are much more likely to recall being in more pain than they actually were, compared to those for whom pain is a fleeting companion. Inaccuracies in the recollection of pain get worse over time. And the longer people have chronic pain, the more depression and anxiety they develop, the more inaccurate their memories of pain get. In one study by Apkarian, 77 percent of chronic back pain patients had memories of pain much greater than what they reported at the time.[38]

Our knack for learning is governed by our ability to form new neurons in the hippocampus—a process called adult hippocampal neurogenesis. Work in mice by Apkarian's lab shows just how our ability to learn, and unlearn, is tied to the transmutation of acute into chronic pain. Apkarian used three different methods to manipulate the hippocampus and showed that reductions in neurogenesis led to either complete blockage or reduction in the development of chronic pain, despite the infliction of a variety of injuries.[39]

Reduced adult hippocampal neurogenesis is also associated with depression. In one experiment, mice with reduced neurogenesis showed more depressive behavior, along with an increase in persistent pain. Subsequent work in patients with chronic back pain by Apkarian showed that the shape of the hippocampus could predict which patients might have exaggerated recall of previously experienced pain. Memories tie into the origin of phantom limb pain as well. And we know that those whose memories are impaired, such as people with dementia, also report less pain.[40]

Yet the clinical definition of chronic pain continues to assume it to be a direct continuation of acute pain, despite the science showing that chronic pain has a distinct footprint in the brain. Chronic pain is as different from acute pain as seeing is from listening. And while acute pain has an obvious evolutionary purpose, chronic pain almost seems like an evolutionary misstep. Yet our burning desire to learn from acute pain may be the reason why its ripples can be so enduring.

Of all the voices in our head, pain is the loudest. And our minds are designed to learn from pain so that we don't repeat what took us down that path in the first place. If we cut ourselves while chopping vegetables, or if sleeping on our bellies gives us a terrible backache, our bodies want us to remember those triggers both as vividly as possible and for as long as possible so that we never rekindle those wounds. Our body would much rather entertain many false alarms about bodily damage than miss a single important one. If our pain system were a court, it would rather put many innocents in prison than let one dangerous criminal loose.

While many assume evolutionary mechanisms to be perfectly designed and glitch-free, reality is more complicated. For one, our feet were initially designed, like those of chimpanzees, to be able to grasp branches; they haven't evolved well enough to be able to best do what they do now, which is work as levers that propel us from one step to another. This in part explains why fractures and sprains in our ankles and feet are so common. This is true of our spines as well, which were shaped to help us climb and move about in trees but now are the leading cause of chronic pain in a world in which we spend most of our time on the ground.

The evolution of pain into an attention-grabbing toddler cannot be explained simply by mechanics. Pain and its memory, just like the ability to walk on two feet, were important adaptations for our survival; yet somehow, as it has spread unchecked, chronic pain appears to have outgrown its utility. And if current trends hold, we are only more likely to hurt and suffer longer and deeper as time goes on.

Our misunderstanding of chronic pain has come at a cataclysmic cost. Two days after I spoke to Martha Lundgren and her husband, David, whose chronic pain prevented him from sitting, Martha was in front of her computer in the midst of a work meeting, when David's alarm went off. "Somebody asked what was that noise in the background," so she went to investigate.

"He was lying in a perfectly normal resting position for him. Except his eyes were very slightly open and he didn't respond to more aggressive shaking and pinching," she said. When she felt for a pulse, she detected "a real rapid heartbeat."

Martha immediately called 911. David still had the breathing mask for sleep apnea he slept with on, and the 911 operator asked Martha to take it off. "They said those weren't normal breath sounds and asked me to start CPR." David was in the midst of a cardiac arrest.

Performing CPR can be extremely traumatic, frequently causing fractured ribs and punctured lungs. "The last thing I did to him was horrible," Martha told me. When the emergency medical crew came, they ascertained he had been dead for a while. "I got to lie with his body after they left. I got to cradle him, hold him closely and embrace him."

David had chronic lower back pain and sleep apnea, but he had no other medical conditions—no heart disease, no diabetes, no smoking gun as to why he would go to sleep and never wake up.

Except for one.

Martha had white hair, clipped tightly down, and wore small earrings. She looked up to the ceiling, blinking rapidly, as both tears and the truth of how he died came to her as we spoke.

David had died of an opioid overdose. For years he had been on a combination of OxyContin and other narcotics. He was on such high doses that a minor miscalculation, an extra pill here or there, could easily be too much.

When Martha told me about David's death, I was shocked. I had spoken to them just two days before, and all three of us talked about how their lives might change in the years ahead. That life together was stolen from them.

Martha had begun to clean things up in the house, but it was not an easy process. "I am not erasing him. It's still full of things that he loved, photographs that he had taken," she told me. To put any final affairs in order, she accessed his email and found an unsent message intended for a friend sitting in his drafts folder:

> I get in melancholic moods myself but I do everything I can to not
> play the victim even though western medicine contributed to my

condition. Feels like a lot of it is grief to me—my old self pretty much died a lengthy death.

David was right. The modern Western approach to chronic pain has the blood of many like David on its hands. Millions of Americans have been prescribed opioids, even as long-term studies suggest opioids have no impact on reducing chronic pain, and the best studies actually show that opioids increase how much pain people feel. Between 1999 and 2017, more than seven hundred thousand Americans died of opioid overdoses, and millions more have become dependent on these drugs.[41]

This misunderstanding of chronic pain was far from an accident. The amorphous and multifaceted experience of pain was hammered into a unidimensional physical sensation by one of the most carefully engineered campaigns of disinformation in human history.

5.

THE GOD OF DREAMS

The History of Opium and the Cultural Transformation of Pain

I think modern medicine has
   become like a prophet offering a
   life free of pain.
It is nonsense.
                    —Elisabeth Kubler-Ross

AMERICANS CONSTITUTE ONLY 5 PERCENT of the world's population but consume 30 percent of the world's opioids. Tens of millions are addicted. Fatal opioid overdoses have killed more than half a million Americans so far, and there's evidence this might be an underestimate. One study suggested that between 1999 and 2016, an additional 99,000 Americans may have died from opioid overdoses that were simply not identified as such. And from April 2020 to April 2021, according to data collected by the Centers for Disease Control and Prevention, more than 100,000 Americans died of drug—mostly opioid—overdoses, a record for any year in American history and double the number of deaths in 2015.[1]

Opioids have infiltrated every facet of Americana, touching the lives of every American whether they know it or not. One might think that the ability of opioids to overpower our wills would be tempered by knowledge of the harms these drugs can cause. Yet some of the very people with the most exposure to narcotics face the greatest risk of succumbing to their bewitching magnetism. Physicians, especially those most likely to prescribe opioid painkillers—anesthesiologists and pain specialists—are highly susceptible to becoming dependent on and overdosing on opioids.

Hospitals rarely, if ever, shut down, and a doctor's work never stops. Yet, even as hospitals remain ever awake, their unblinking gaze permanently pried open with A *Clockwork Orange*–style eye clamps, day and night perpetuate a ceaseless and necessary cycle of renewal. Though hospitals become quieter by night, they are overrun by a veritable army of staff cleaning every nook and cranny. Their labor may seem procedural, but it is critical to the ecosystem of the hospital and its inhabitants. The slow-moving scrubbing machines that patrol hallways pulverize not just bacteria and bodily fluids but the horrors and tragedies that take place in operating rooms and hospital wards. Such vigorous and constant rejuvenation is necessary not just to keep the surfaces gleaming and aseptic but to prop up the spirits of those doctors and nurses charged with the preservation of life and dignity. Not all stains can be erased, though, not all specters subsumed. And one such specter was the ghost of Brent Cambron.

No one talked about Cambron when I joined the hospital where he reached rarefied heights of success. Valedictorian of his high school class in tiny Sperry, Oklahoma, Cambron was preternaturally intelligent and left home after medical school to join the Beth Israel Deaconess Medical Center in Boston as a resident in its prestigious anesthesiology department, where I worked as a research fellow. His hands were the coolest on deck whenever a patient crashed; his leadership abilities were recognized with his selection as chief resident by his peers and faculty.

Yet his buoyant rise was punctured by a single pinprick when he injected morphine that he had stolen from work into his muscle. While initially he was able to maintain his habit covertly as he finished his residency, specialized in pain medicine, and was hired to join the faculty at Beth Israel, it finally caught up with him, as evidence of delinquency and personal and professional neglect accumulated, leading to his dismissal and suspension of his medical license in June 2007.

His addiction wasn't done with unravelling his life just yet. It pulled him back to the hospital even after he had been fired, where he would sneak in at night using scrubs as his disguise, filling backpacks full of syringes and vials. Police arrests and rehab stints didn't deter him until one day in October 2008, in a storage closet in the hospital that we both trained at, he was found lifeless, dead at the hands of the very drugs he was trained to master.

For all the promise and potential that ended when Cambron's body was discovered that morning, his memory was minimized to an inaudible whisper crisscrossing the place of work that became his final, unmarked resting place. Human beings are no match for the vexatious grip of opioids, no matter how experienced they are at their administration, how mindful they are of their mesmerism.

In coming to terms with the cause of the opioid crisis, much attention has been focused on Purdue Pharma, the manufacturer of OxyContin. But the fact is that within the medical community they found abettors and enablers at every turn. From physicians to pharmacists, from medical schools to research laboratories, from consulting firms to distributors, from regulators to lawmakers, all fell over themselves to accept the Sacklers' money and advance their agenda. It was just too easy.

What allowed opioids to be prescribed so widely was the epidemic of chronic pain that they were supposed to be quashing. But they achieved the opposite.

Far from reducing the number of people in pain, opioids increased how many Americans suffer from debilitating discomfort.

Far from helping us better understand chronic pain, opioids erased whatever little we knew about the nature of suffering.

Far from making the management of pain equitable, opioids exaggerated the biases that lead to vulnerable people's agony going unattended.

Far from bringing patients and physicians closer, opioids pitted them against each other, beginning with indiscriminate overprescription, followed by abrupt cessation, turning the clinic visit into a face-off.

Almost every basis upon which the opioid epidemic was built was a carefully scripted lie, perhaps none more consequential than one pushed by one of medicine's most respected organizations: that one hundred million Americans were living with chronic pain.[2]

The opioid epidemic has been a true catastrophe, but it is not a contemporary one. Our modern epidemic was hundreds of years in the making. To really understand how millions of Americans came to be dependent on opioids, we have to trace the connection between opioids and chronic pain and place it in its right context within the arc of human history.

By now, many people know the origin story of the opioid epidemic.

The nineties were a time when pain was rampant in America. A movement had grown that characterized pain management as one of medicine's biggest blind spots, but new, reportedly less addictive narcotics could change that. Opioids, historically taboo treatments, proved to be one of the few remedies in a physician's toolkit that could quickly quiet pain. Around this time, new supplies from countries like China fueled even greater demand and rampant abuse, with one in two hundred Americans becoming addicted to opioids. Sensing a growing problem, authorities clamped down, and many users turned to even more potent illegally acquired opiates, making a swift death only a miscalculation away. Physicians eventually took responsibility, championing a more limited role for opioids. Pharmaceutical companies developed new opioids. Again, these were sold to the public with promises of lower risk for abuse. Again, those claims were lies, as these new opioids turned out to be even more lethal than the ones before.

These, in fact, were the *eighteen* nineties, an eerie echo of events that would repeat again a century later in the nineteen nineties.

In focusing on the misuse of prescription opioids today, it is easy to forget that opium, the substance from which they are derived, has a long and important history of its own. But the story of the poppy always takes a predictable cycle in human society: it starts off as our savior and ends as our scourge.

Opium may well be our oldest known drug. The first evidence of the cultivation of opium, derived from the poppy, was inscribed on stone tablets by the ancient Sumerian people almost eight thousand years ago. They called it *hul gil*, the "plant of joy." The Sumerians passed down their knowledge of the poppy to the Assyrians, who gave it to the Egyptians, who then spread it around the known inhabited world. Opium has since been a part of almost every civilization.[3]

In ancient Greek culture, the poppy was elevated to divine status. It was often used in religious ceremonies, and ancient figurines called poppy goddesses have been found across the region. Modern medicine can trace its roots to ancient Greece, where we find the first recorded use of opium as a medicine rather than just a drug. For that, we have the pioneering Greek physician Hippocrates (460–370 BC) to thank.[4]

Hippocrates wrote at great length of the poppy, differentiating the white, black, and fire-red varieties, and was able to acquire various effects from it depending on whether it was ripe, unripe, or baked. Opium was used as a sedative, as a means to slow diarrhea, and as a treatment for aches and pains. Even when it was first brought within the fold of medicine, its addictive potential and toxicity appeared to be well-known. The philosopher and astronomer Heraclides Ponticus (390–310 BC) wrote about how opium was often used to achieve what some regarded as a good death. In "On Government," he wrote of the inhabitants of the island of Keo who "do not wait until they are very old for death to take them, but before they grow weak or disabled in any way, take themselves out of life, some by means of the poppy, others with hemlock."[5]

As its use became widespread, many of Hippocrates's disciples began to push back on opium's medicinal application. Epistratos claimed that opium "dulled the sight and is a narcotic." Galen, the Roman physician who stands next to Hippocrates as one of the most influential physicians of all time, called it the "strongest of drugs which numbs the senses and induces a deadening sleep." An account in his writings strongly suggests that Marcus Aurelius, the Roman emperor and Stoic philosopher, may have been addicted to opium: Galen used to prescribe a concoction to help him fall asleep. When Marcus Aurelius asked Galen to remove the opium from it, the concoction stopped working, and insomnia brought him back to Galen, insisting on the readdition of opium.[6]

Throughout history, opium has been deeply symbolic, often related to the different ways it was used. In antiquity, when the poppy emblazoned the crowns of entranced priestesses, it was a symbol of fertility and harvest. In ancient Greece, the poppy was the symbol of Demeter, the goddess of agriculture, who is purported to have discovered the poppy.

As opium's medicinal uses became more common, the poppy came to be associated with healing. Yet, as the use of opium follows its natural life cycle, beginning often as a sign of affluence and healing, in time it inevitably transforms into a symbol of death. In Greek mythology, Thanatos, the god of death, is depicted carrying a wreath of poppies. The symbolic turn of the poppy is also seen in the story of Demeter, who overdoses on the milk of the poppy to induce anesthesia and forget the torment of having had her daughter raped and abducted by Hades. Poppies, associated with a swift and painless death, were often carved into gravestones.

This symbolic metamorphosis of opium from healing medicine to lethal toxin has occurred again and again throughout human history. Almost every patch of land on the planet has been deeply touched by opioids, but South Asia has a particularly long history with them. While initial reports suggested Arabs introduced opium to the Indian subcontinent, more rigorously performed archaeological research suggests that it was Alexander the Great, who brought with his invading armies a rich supply of opium.

By the sixteenth century, opium was commonly used on the sub-continent, and one of its most voracious connoisseurs was the Mughal emperor Jahangir. Jahangir, whose name means "world seizer," inherited one of the greatest empires ever after the death of his father, Akbar the Great. His son would commission the Taj Mahal. Yet Jahangir himself became so addicted to opium that he could barely rule his kingdom, essentially abdicating and leaving his wife, Nur Jahan, to lead on his behalf.[7]

Opium was Jahangir's undoing and perhaps the Mughal Empire's as well. It never reached the same heights as it had during the reign of his father. As the Mughal Empire weakened, it was eventually usurped by the British East India Company, not just the world's first global corporate giant but the most successful drug cartel of all time.[8]

Looking closely, one can see how closely Purdue Pharma's strategies followed the plans laid out by European imperialists more than a century earlier. As one falls, another rises, and the cycle repeats.

In the eighteenth century, the British faced a dilemma. They had developed an insatiable taste for tea but lacked anything they could sell in return to pay for it. China was a huge potential market, but what could the British possibly sell them? "We possess all things," the Qianlong emperor wrote to George III, "and have no use for your country's manufactures."[9]

The British decided to change that. The East India Company forced the farmers and peasants of the Indian subcontinent to grow opium on an industrial scale previously unseen in human history. Not only did they sell the drug back to the same South Asians whose toil they had monetized, but they aggressively used smugglers to infiltrate China and create a thriving market for opium there. While normal economic principles state that increased supply can be detrimental, opium's highly addictive nature turned the tables, with excess supply fueling its own demand. This strategy was extremely effective: Chinese opium smokers spent more than twice on their habit than the entire budget of the Chinese government.[10]

In response, the emperor banned the recreational use of opium in 1799, but the British remained undeterred. Matters reached a crescendo in 1839, when Chinese authorities seized more than a thousand tons of opium held by mostly British smugglers. At the behest of the smugglers, the British government demanded the full value of the opium back. When the emperor refused, the British sent in their warships, and thus began the First Opium War. The war ended in 1842 with Britain emerging as the handy victor and seizing Hong Kong, which it would retain for another 155 years. The British would go on to wage several additional wars, forcing China to legalize the import of opium. At the same time, the sale of opium remained banned on British shores—unless the sale was to someone of Chinese or Indian origin.

The Opium Wars shape the world to this day. Many Chinese still use them as a prism for viewing the West. Yet, in our time, the roles have become more complicated. China today is the world's chief manufacturer of fentanyl, a synthetic opioid far more potent than morphine. Between 2013 and 2018, fentanyl and similar drugs contributed to a tenfold increase in overdose deaths in the United States. And while the United States is pressuring China to clamp down on production and Beijing has taken some steps to do so, the Chinese government also responds that the problem is the astronomical demand for fentanyl in the United States. While much has changed since the nineteenth century, one economic principle hasn't: the oversupply of opium creates its own insatiable demand.[11]

Given all this history, it is remarkable that white Europeans continued to so closely associate illicit drugs with people of color. In his book *Opium and Empire*, historian Carl Trocki writes, "For European observers, one of the most enduring nineteenth-century images of the Chinese...was that of the opium 'wreck.' The hollow-eyed, emaciated Oriental stretched out on his pallet, pipe in hand, stood as the stereotype of Asiatic decadence and indulgence...The victim had come to stand for the crime, and the image has acquired an extraordinary historical durability."[12]

Even as the British ran their global drug cartel, they portrayed their drugged subjects as the real savages. Such is the privilege of white supremacy. "A modern empire...is the most efficient instrument of comprehensive reforms in law and government and the most powerful engine whereby one confessedly superior race can control and lead other races left without nationality or a working social organization," wrote Alfred Lyall, a British bureaucrat who rose to home secretary of the British Indian government. Such racist tones to this day affect whether we call someone who uses opioids an addict or a felon, whether we treat addiction as a crime or a disease, all based on the color of the user's skin.[13]

The modern opioid epidemic, like its nineteenth-century counterpart, didn't come into being through an inopportune alignment of circumstances. The opioid epidemic of our times was one of the most carefully orchestrated, intentionally designed calamities in the history of mankind.

Even as eerily similar cycles of opioid outbreaks hit various societies throughout history, not each cycle was the same. The use and eventual misuse of opioids were always tied to wider social and economic currents. This was particularly true of the opioid epidemic that struck one of the youngest British colonies in the eighteenth century: America. The discovery of morphine and the hypodermic needle meant that the pushers of this American plague were not smugglers but physicians and pharmacists.

Here, in the wake of America's first great war, lay the true inception of today's opioid crisis.

Throughout the course of history, when a person experienced debility, they went to a mosque, spoke to a rabbi, or prayed for a miracle from the heavens. They shared their anguish with their families and friends in the communities they lived in. Pain was not just the result of receptors firing in one's body but a metaphysical conundrum that required resolution. America's opioid problem was unique because it coincided with the rise of modern medicine, not just as a scientific specialty but as a philosophy and way of life. What you felt in your body was a biophysical kink rather than a spiritual poke. Pain, which always held supernatural meaning, was now nothing more than a physiological wrinkle needing to be ironed out.

Medicine changed how we lived with the agony arising within. It encouraged action rather than stoicism, prescriptions rather than prayers. And America was patient zero in this grand experiment.

Fought between 1861 and 1865, the American Civil War led to an estimated 750,000 deaths, exceeding the toll of all other American wars combined. The deaths were sadly just the tip of the iceberg. The minié balls that tore through flesh and bone so effectively led to grievous injury and often amputation. Field hospitals quickly became littered with piles of dead limbs.[14]

The American Civil War was not the first brutal conflict. But it was the first to be waged after the Industrial Revolution and the widespread availability of an instrument sometimes regarded as the greatest medical innovation of all time: the hypodermic syringe.

The use of opium had disappeared from Western Europe during the Dark Ages, but it surreptitiously made a comeback in the form of laudanum, a tincture of opium mixed with alcohol. Opium began to be administered in every which way—eaten, inhaled, rubbed, even inserted rectally—for all sorts of conditions, including pain, asthma, alcohol-induced seizures, diarrhea, menstrual cramps, morning sickness, and teething. Yet the substance and its preparation remained largely unchanged from when the ancient Sumerians first processed it thousands of years ago.

In 1805, a young German pharmacist, Friedrich Wilhelm Sertürner (1783–1841), became the first person to isolate opium's active narcotic substance. He named the purified white crystal after

Morpheus, the Greek god of dreams. After experimenting on dogs (at least one of whom died), three other young men, and, as was the norm then, himself, he finally published his findings in 1817. The isolation of morphine, called "the world's first true drug," represented a massive advance over prior formulations because of how precise one could be about how much active substance was being administered. Previously there was no way of knowing how potent two different opium products would be, even if they were of the same brand and purportedly contained the same dose.[15]

Yet nobody really knew what to do with morphine until an enterprising Scottish physician devised a method to inject it straight into the human body. His innovation, modeled after the sting of the wasp, was the hypodermic syringe.

Alexander Wood had been called to help treat severe nerve pain that an older woman had been having around her collar-bone. She had been unable to sleep for three nights straight. However, this woman couldn't ingest opium, which had previously caused her to faint. On November 28, 1853, Wood went to her home at 10 p.m. "to give the opiate the benefit of the night." The twenty drops of morphine he injected into her shoulder would mark the first time an opioid was injected directly into the body. "In about ten minutes after the withdrawal of the syringe the patient began to complain of giddiness and confusion of ideas," he wrote in his historic paper in 1855; "in half an hour the pain had subsided, and I left her in the anticipation of a refreshing sleep."[16]

When he came back the following morning at 11 a.m., she was still comatose. "[I] was a little annoyed to find that she had never wakened; the breathing also was somewhat deep, and she was roused with difficulty."

Apparently her nerve pain never returned.

Alexander Wood's case series went on to describe the mostly female patients on whom he experimented by injecting morphine right into the painful part of the body using his custom-built syringe. His paper was ridden with the casual sexism and classism characteristic of that time. One thirty-year-old woman was short and of "plump habit" and "suffer[ed] from flatulence"; one fifty-year-old widow had a "hysterical temperament," while a twenty-three-year-old woman was "in the lower rank of life."

While Wood initially hypothesized that the injected morphine would have an effect only on the local tissues, after he had seen a number of cases of intoxication, coma, hallucinations, and other body-wide effects, he concluded that "the effect of narcotics so applied [was] not confined to their local action, but that they reach[ed] the brain through the venous circulation, and there produce[d] their remote effects."[17]

The hypodermic needle spread all around the world. In England, it became a fashionable accessory for high-society elites, with many women carrying personalized, bejeweled syringes. It was a symbol of sophistication and affluence, a phallic, almost sexual symbol that made its way into novels such as Bram Stoker's *Dracula* and the opening passage of Arthur Conan Doyle's *The Sign of Four*:

> Sherlock Holmes took his bottle from the corner of the mantel-piece, and his hypodermic syringe from its neat morocco case. With his long, white, nervous fingers he adjusted the delicate needle and rolled back his left shirtcuff. For some little time his eyes rested thoughtfully upon the sinewy forearm and wrist, all dotted and scarred with innumerable puncture-marks. Finally, he thrust the sharp point home, pressed down the tiny piston, and sank back into the velvet-lined armchair with a long sigh of satisfaction.

At the same time as the hypodermic needle gained cultural cachet, other forms of ingesting opium, such as inhalation, became increasingly taboo. The reasons were entirely racist and xenophobic: opium smoking was strongly associated with Chinese and other immigrants.

Yet it was among white Americans that morphine and the syringe showed their potential to engender both instantaneous relief and life-long agony.

Morphine quickly made its way into battlefront first aid kits and field hospitals. Millions of opium pills and tinctures were prescribed during the Civil War, unleashing America's first opioid epidemic. "Even if a disabled soldier survived the war without becoming addicted, there was a good chance he would later meet up with a hypodermic-wielding physician," writes David Courtwright in *Dark Paradise: A History of Opiate Addiction in America*. "Though it [morphine] could cure little, it could relieve anything." Opioid addiction quickly crossed over from veterans of the war to the public, with a majority of addicts being white, upper-middle-class women. In 1888, opioids were the most commonly prescribed drug. Almost one in two hundred Americans became addicted, three times more than in the twenty-first century.[18]

In 1898, in the face of rising concerns about opioid addiction, the German pharmaceutical company Bayer developed a new formulation that was advertised as less habit-forming. A physician wrote in the *New England Journal of Medicine* in 1900 that it was "not a hypnotic" and that it carried no "danger of forming the habit." Diacetylmorphine, however, is better known by the trade name it was sold under—heroin—derived from the German word *heroisch*, meaning heroic. Heroin was prescribed heavily for years before it was revealed to be even more addictive than morphine.[19]

Nobody saw this as much of a problem until Chinese immigrants began arriving on American shores and opioid use proliferated among low-wage workers. Now it became much easier to pass draconian laws to outlaw not just illicitly acquired opioids but clinics that were giving controlled doses of opioids to manage people's pain or addiction.

The criminalization of opioids, however, only shoved them into the shadows. Many who had been taking prescription drugs turned to illegally acquired heroin and morphine. As attitudes toward opioids changed, so did how we live in pain.

The Industrial Revolution fundamentally shifted how we relate to our own body, which was now seen as a mechanical appliance, with pain being a beeping alarm that sounded when the gears started to grind. And morphine became the lubricating grease that you could pour over the cogs to get the body rolling again.

The evolution of medicine's new antiseptic approach to the management of pain is captured in the stories of two men of the same name. The first, Ivan Ilych, is the titular subject of a novella by Leo Tolstoy published in 1886. And from the first moment a reader encounters him, they know how his journey will end. Tolstoy named his novella *The Death of Ivan Ilych* because he wanted readers to know the character's destination.

The story of Ivan Ilych is not about where he is going. It is about the road.

Ilych is a forty-something-year-old judge in Saint Petersburg living the bourgeois life until one day he falls and develops pain in his abdomen. Instead of taking the obvious course that most such pains do—spiking sharply after the initial insult, followed by a gradual decline until all memory of it is lost—this pain sticks around, causing Ilych to consult a physician.

For Ilych the most important questions are primal: Is the pain serious? Will he ever be the same again? Is he going to die? Yet his doctor is more focused on the technical elements of the pain and whether it is from "a floating kidney, chronic catarrh or appendicitis."

Far from helping, the doctor's aloofness triggers the slow explosion of Ilych's inner world: "And this pain, this dull gnawing ache that never ceased for a second, seemed, when taken in conjunction with the doctor's enigmatical utterances, to have acquired a fresh and far more serious significance. With a new sense of misery, Ivan Ilych paid constant heed to it."

As pain evolves into suffering, it forces us to ask tough questions of it and of ourselves. We have been wired not just to feel pain but to figure out its cause so that we can learn from it. Therefore, the pain that lends itself most to filling us with dread is not that which accompanies a gaping, bleeding wound or a grapefruit-like tumor in our belly. It is pain that violates the cardinal rule of pain wired into the brains of human and animal alike: it is a pain without reason.

Throughout Ilych's story, the cause of his pain not only remains undiagnosed but becomes irrelevant. A rigorously regimented physician will find it maddening to read a patient's account without knowing what ails them. Yet this is exactly how doctors fail patients like Ilych, who care much more what that ailment means to them.

"You know perfectly well you can do nothing to help me," Ilych tells his doctor, "so leave me alone."

"We can ease your suffering," the doctor replies.

"You can't even do that."

His doctors have access to the most cutting-edge painkillers of the nineteenth century. "They gave him opium and injections of morphine, but this did not relieve him."

Ivan Ilych's death from his unnamed malady is preceded by his "screaming desperately and flailing his arms" for several hours. Yet, within him, an epiphany occurs. After struggling with his pain for so long, he has finally given himself away to it. Before he dies, even though he is in pain, he is not suffering. "There was no fear because there was no death," writes Tolstoy as he ends Ilych's tale with an unexpectedly optimistic flourish.

"In place of death there was light."

In Ivan Ilych's story, Tolstoy presents one of the richest palettes of human suffering in literature. Tolstoy frowned upon a death dulled by religious zeal or opiate analgesia, and he gave Ilych a death many in modern times might find inhumane. Tolstoy appears to concede that bodily pain can relieve us of a far more excruciating feeling native to all sentient beings, guiding our every twitch since our animation—the existential dread of our demise.

Perhaps nowhere would the seismic shift in modern science reverberate more powerfully than in the body of the man or woman in pain. And no one chronicled this brave new world more vividly than a man whose name also happened to be Ivan Illich (1926–2002). Born in Austria, Illich was a priest and philosopher who found his calling in opposing the ill effects of modern technology and societal reconstruction that stunted human values and culture. He lamented the institutionalization of mass education but reserved some of his most biting criticism for mass medicine in his incandescent 1976 book *Medical Nemesis: The Expropriation of Health*. Wielding his pen like a flamethrower, Ivan Illich set the medical world ablaze, asking questions of how we treat pain and suffering that are even more relevant today than in his time. A reviewer in the *New York Times* wrote, "No polemicist writing today has his passion, his range, his glittering and pyrotechnic arsenal."[20]

In the mythological Greek world, the greatest sin was hubris, and Nemesis was the goddess of divine retribution, out to put humanity in its rightful place, prostrate in front of the gods.

As Illich saw it, the greatest sin of the modern world was rampant, massive industrialization, particularly in medicine. In his view Nemesis had come back to earth to punish us for our egotism with her sharpest dagger yet: iatrogenesis, a harm or disease state wrought by medicine itself. "Iatrogenesis is clinical when pain, sickness and death result from medical care; it is social when health policies reinforce an industrial organization that generates ill health; it is cultural and symbolic when medically sponsored behavior and delusions restrict the vital autonomy of people by undermining their competence in growing up, caring for each other, and aging, or when medical intervention cripples personal responses to pain, disability, impairment, anguish and death," he wrote.

Illich reserved his fieriest wrath for how medical civilization turned pain into a technical matter in its quest to conquer it: "Culture makes pain tolerable by integrating it into a meaningful setting. Cosmopolitan civilization detaches pain from any subjective or intersubjective context in order to annihilate it."

Society transitioned from viewing everything through a moral lens, with pain seen as an evil or a challenge to overcome with rituals, prayers, and narratives of honor and duty, to considering pain as something not to be borne but to be quashed. "Pain thus turns into a demand for more drugs, hospitals, medical services, and other outputs of corporate, impersonal care... [giving] rise to a snowballing demand on the part of anesthesia consumers for artificially induced insensibility, unawareness, and even unconsciousness."

The modern world deemed pain intolerable, discomfort an unacceptable state. And medicine placed itself as the only legitimate source of relief. This, in turn, allowed it to fortify its position in contemporary capitalistic society.

Even Ivan Illich's worst prognostication about the industrialization of medicine might not have foreseen just what the last few decades have witnessed. Healthcare accounted for 5 percent of US GDP in 1960, tripling to 18 percent in 2016. This astronomical growth has not led to better health; if anything, the United States lags other industrialized countries in almost all health metrics. Furthermore, the American healthcare system has become an instrument of inequity and injustice, deepening the fissures between the haves and the have-nots.[21]

Far from alleviating suffering, modern healthcare is designed to make us hurt more and more.

More than helping people find relief, modern healthcare has figured out how to monetize the poorest and most pained people in the United States. Just 5 percent of Americans account for half of all US healthcare spending. This money is not spent preventing people from getting sick in the first place. It is not spent helping many of these high-need patients feel better or live the lives they would like to. It does, however, enrich a trillion-dollar healthcare system, which spends more on administrative costs than any other nation on earth.[22]

This expansion of modern medicine was never just about building more hospitals and clinics. Modern medicine has also crafted a philosophy supplanting millennia of cultural norms and positioned pain as a purely physical sensation that only medical interventions can alleviate. Ivan Illich was one of the first few to ring the alarm on the coming wrath that the gods would bring down on a species dreaming of an existence free of pain and suffering.

This surging desire for doctors to rid humanity of suffering was coupled with the emergence of a biomedical, pharmaceutical industry now worth a trillion dollars and promising mass anesthesia. The pharmaceutical industry not only created products that could treat myriad conditions and symptoms but reshaped human society fundamentally.[23]

The confluence of advances in medicine and a barrage of pharmaceutical companies marketing directly to doctors and patients birthed a movement that deemed suffering unacceptable. As spiritual voids gaped inside those economically left behind and loneliness became a way of life, people in pain kept being sent exclusively to doctors' offices and pharmacies. The drugs of choice to treat pain happened to be the most lethal compounds ever to be mass-ingested by human beings.

Every single opium high follows a somewhat predictable cycle, with the initial euphoric rush followed by peace and comfort; then the effect wears off, and the pangs of withdrawal take hold. After he discovered morphine, Friedrich Sertürner began to use it on himself, and soon thereafter, as the full spectrum of the drug's effects manifested themselves, he came to view the discovery that would immortalize him differently. "I consider it my duty to attract attention to the terrible effects of this new substance," he wrote prophetically in 1810, "in order that calamity may be averted."[24]

Sertürner's personal journey ricochets throughout the personal biographies of those in thrall of opium's promise. The Arab polymath Avicenna, who pioneered the use of opium during the golden age of Islam, when Western Europe was languishing in the Dark Ages, himself died of an opioid overdose. The wife of Alexander Wood, the man who first injected morphine using a syringe, was widely rumored to be a morphine addict, with some claiming she died of a morphine overdose. And closer to me, the same drug that Brent Cambron had seen provide comfort to so many in his care enkindled his own annihilation.

Yet societies have also experienced opioids in similar cyclical loops since time immemorial. Just as physicians welcomed morphine as a panacea for all our ills in the 1800s, in the 1990s that entire story would repeat with the introduction of a new compound: OxyContin. Like every previous innovation in opioids, it was deemed to be both more effective for pain relief and less addictive. Yet, even as the history of opioids is cyclical, our response too has always been the same: we succumb hopelessly every single time, learning nothing that might help prevent future cycles. It is time, however, for us to break that cycle once and for all and get to the heart of what gave rise to our current calamity.

The spread of opium has often been associated with larger cultural and political inflection points. Colonialism pushed opioids onto groups not just to subjugate them but to use their addictive qualities to further discriminate against and criminalize people of color, an attitude that persists to this day. The development of morphine and its widespread use in the Civil War led not just to the industrialization of opioids but to a transformation in the cultural history of pain itself. The secularization of pain drained the most complex sensation we feel of all meaning, while the surging biomedical industry advertised eternal numbness as a desirable and attainable state. Yet, in the twentieth century, especially after World War II, medicine and opium would become tied to the third major inflection point at the heart of the opioid epidemic: the rise of consumerism.

6.

ANGEL OF MERCY

How We Learned to Stop Worrying and Love the Pill

We are healed of a suffering only by
   experiencing it to the full.
                —Marcel Proust

AFTER FINISHING UP MEDICAL SCHOOL in Pakistan and moving to the United States, I was prepared for everything that a sick person might ask of me—except for opioids. And I would learn how to prescribe them not from a textbook but from my patients themselves.

The medical school I attended in Pakistan is known for producing some of the best-trained physicians in the world. But opioids were simply not part of the curriculum because they were rarely used. They were prescribed almost exclusively for patients who had just undergone a major operation or sustained some horrific injury.

When I started my internship in the United States in 2011, on my very first day a patient in the hospital asked me to prescribe him Dilaudid, the brand name for hydromorphone, a very potent opioid. I told him I had never heard of it. He then spent the next few minutes explaining the pharmacology of Dilaudid to me as I stood at the foot of his bed. He told me how quickly it took effect, how long it lasted, and when he would need his next dose. He told me it only worked when it was given quickly, with a push, rather than as a slower infusion. This humbling experience foreshadowed a truth that I only now recognize: the opioid crisis would come to define the medical training of my cohort of doctors around the United States just as HIV/AIDS had before us and COVID-19 has after.

We were the opioid generation.

Opioids marked every day of my internship. I remember a patient, a lawyer from New York, who came into town for a meeting but had a flare-up of his chronic pancreatitis. We asked him to stop eating, since ingesting food can make the pancreatitis worse, and gave him painkillers through an IV. However, we soon realized everything he told us was a lie. He wasn't a lawyer from out of town, and he didn't have pancreatitis. He was going from one Boston hospital to another, repeating the story, getting more IV painkillers to get high, and then, once discovered, moving on to the next. An endoscopy revealed that his stomach was full of food he had been eating secretly throughout.

Another woman came to our hospital with symptoms consistent with superior vena cava syndrome, in which blockages of veins in the chest can lead to swelling of the face. It turned out she just had a large face and had been using that to dupe doctors into giving her powerful painkillers for years. We caught another patient in the emergency room adding drops of blood from his fingertip into a urine sample to simulate the effects of a kidney stone in order to receive opioids often prescribed for those experiencing this spine-splitting ordeal.

It is hard to overstate how damaging such incidents are to a young doctor's psyche. We go to medical school to learn how to take care of those who come into the hospital, not to sniff out cons. Yet it has become common for doctors and nurses to google patients to make sure they are telling the truth about themselves.

At the same time, we were overwhelmed with patients suffering adverse effects of opioids. I once took care of a young hospice nurse who overdosed on morphine that she had been stealing from her terminally ill patients. She had severe brain inflammation and never woke up from her coma. One patient had such severe constipation from opioid use that his stool tore through his intestines. He died shortly afterward. I learned all I know about diagnosing brain death from the young men and women coming into the hospital after overdosing on opioids.

Almost everything we ever do in medicine involves serious risks. Even something as benign as a baby aspirin can cause a brain bleed. But opioids are in a risk class of their own. "I had a patient die," one nurse revealed. "He took the entire bottle, and the police came to see me because they found him dead with the empty bottle with my name on it." Shaken by the incident, she now tells patients when she writes them a prescription, "I don't want you found dead with my name on your bottle."[1]

An analysis of almost thirty-seven hundred patient episodes in the hospital shows that opioids are responsible for a quarter of all adverse drug reactions—ranging from confusion, constipation, drowsiness, and nausea to life-threatening accidental overdoses requiring the administration of the opioid reversal agent naloxone. Often when I am called in the middle of the night for a patient who has stopped breathing, I reflexively reach for naloxone because of how common overdoses are in the hospital.[2]

Between those faking symptoms to receive opioids and those suffering from their complications were the majority with legitimate chronic pain, which at times seemed to be each and every one of my patients. In theory, the benefits opioids offer are supposed to make their risks palatable, particularly for those who suffer chronic pain. Yet patients with chronic pain who receive opioids in the hospital are actually less likely to feel better from them than patients without chronic pain who receive opioids. Patients prescribed opioids in the hospital also stay longer and fare worse when they leave: they are more likely to be readmitted, to have a poor quality of life, and to have many more visits to the clinic and emergency room than those prescribed other painkillers, such as anti-inflammatories.[3]

An extensive analysis of qualitative studies backs up my own experiences. Opioid prescription requires clinicians to tiptoe on moral boundaries as they thread the needle between relieving and compounding suffering, walking a fine line between treating pain in the present and preventing pain in the future, responding to the cries of the patient in front of them and guarding the community they live in.[4]

Opioids have infiltrated every aspect of medical culture. One of the most pervasive aspects of medicine is mnemonics, which help us memorize often complex concepts. One of the most famous ones is MONA, for the treatment of people with chest pain potentially due to a heart attack. The M stands for morphine (O is for oxygen, N for nitrates, and A for aspirin). Yet people who receive morphine are more likely to have heart attacks and to have more massive heart attacks when they do, and they do not absorb critical medications that help treat the heart attack. Despite this, partly because of the pervasiveness of this mnemonic and partly because of American medicine's propensity to hand out opioids like candy on Halloween, most people who have heart attacks are still given morphine by emergency medical staff. In fact, a patient of mine who was in the intensive care unit threatened to leave the hospital in the midst of having a heart attack if I didn't give him more morphine for his chest pain.[5]

It was around this time, in 2011, that I read a congressionally commissioned report by the Institute of Medicine, *Relieving Pain in America*. Since renamed the National Academy of Medicine, this prestigious nonprofit institution, per its mission statement, provides "independent, authoritative, and trusted advice nationally and globally." I was hoping that the 364-page document might offer strategies that I could use to balance managing people's physical discomfort with the dangerous toxicity of opioids. In fact, the report simply warned against restricting their use.[6]

A few days later I would learn about a glaring omission in the pain report: nine of the nineteen experts involved had strong financial connections to or served on the boards of companies that manufactured opioids or worked for groups that were under federal investigation for conflicts of interest, according to an investigation published in the

Milwaukee Journal-Sentinel. *Relieving Pain in America* was responsible for introducing one of the most influential falsehoods in the annals of pain: that one hundred million Americans suffered from chronic pain, a figure disputed by the very researcher whose work the number was derived from. Even the origin of the report was tainted: many of the lawmakers who had asked for it had received hundreds of thousands of dollars from opiate manufacturers.[7]

Learning all this shattered the idealistic view I, a previously bright-eyed intern, had of my profession. Despite its noble aspirations, American medicine and academia had served as witting and unwitting accomplices in a scheme that addicted millions of Americans to illicit drugs, killing more Americans than World War II. The troubling truth is that despite this tragedy, which is still unfolding, most of medicine has averted its eyes and strolled along without any introspection or reform.

In the middle of my intense shifts in the hospital during my residency, the daily noon conference was an oasis. No screeching pages, medical emergencies, or mountains of paperwork. Just an uninterrupted hour of teaching. We medical trainees sat on chairs in a large classroom and discussed interesting cases, learned critical factoids, indulged in gallows humor, and, perhaps most importantly, put food in our restive bellies. It was a daily reminder of why I wanted to be a physician.

When I look back, those conferences blur together—except for one. Instead of diving into another medical mystery, one of our chief medical residents, John Mafi, read out a note that a recent graduate of the program had sent him. His name was Kevin Selby, and he had moved to Lausanne, Switzerland, where he was completing training as a primary care physician.

John read the email aloud to us as we continued to mow down our lunches (I want to say it was burrito day). It wasn't really clear to anyone why John decided to share the email until he reached the very last thing Kevin wrote to him, something so outlandish and fantastical that it silenced the room. Kevin wrote to John that in his six months of clinical practice since leaving the United States, he had not prescribed a single opioid.

Like many of my colleagues, I could barely go a few hours without prescribing opioids. It was my understanding that I couldn't actually perform my job if I couldn't manage people's pain. I had been trained alongside the rest of the doctors in this country that the only way to help those in pain was to offer them an opioid; yet Kevin's account seemed to challenge that notion.

I recently asked Kevin about this when I caught up with him. "The goal was zero pain. The culture was that you hadn't really taken anyone's pain seriously if you hadn't given them opioids," he recalled about his time in Boston over the phone from Switzerland.

Kevin felt even more pressured when he was working in the emergency room rather than the clinics or wards. "We were told that we had to treat pain fast, that it was unacceptable for people to be in pain and the way to treat pain fast was to give opioids."

This was very different from his experience in Lausanne. But he wasn't seeing fewer people with chronic pain. Studies show that Americans are not much more likely to have chronic pain than Europeans. The difference was how European doctors were trained to respond to it.[8]

As a resident, it seemed as though not being able to give opioids for chronic pain was like playing baseball without a bat. As I would later learn, though, that was because I had been asked to play a losing game. In fact, almost everything I had been taught about pain and opioids was wrong.

In 1895, an estimated three hundred thousand Americans were addicted to opioids largely prescribed by physicians. This was a dramatic shift from earlier in the nineteenth century, when the majority of users were Chinese immigrant workers who smoked opium. Hypodermic needles were advertised in Sears catalogs, and users were mostly upper-middle-class, female, and white. Concerns about rampant drug abuse swelled nationally.[9]

In 1908, Theodore Roosevelt appointed Dr. Hamilton Wright as the US opium commissioner. Wright was a medical doctor born in Cleveland, Ohio, whose work took him to India, China, and Malaysia, to name a few places. His travels exposed him to the dangers of opium abuse around the world, imbuing in him a special zeal against the poppy. He told the *New York Times* that year that thousands of people were "slaves to the opium habit, about five-sixths of whom are white." The Smoking Opium Exclusion Act, passed in 1909, banned the possession, importation, and use of opium for smoking. However, exceptions were made for the "medical" use of opium, which continued to be rampant. In 1911, in another interview for the Gray Lady, he proclaimed, "Of all the nations of the world, the United States consumes the most habit-forming drugs per capita. Opium, the most pernicious drug known to humanity, is surrounded, in this country, with far fewer safeguards than any nation."[10]

Known as the "Opium Doctor," Wright was called an "important leader in a worldwide crusade," and he came down hardest on his own colleagues' role in spreading the plague. "Our physicians use [opium] recklessly in remedies and thus become responsible for making numberless 'dope fiends'... [H]ere physicians often are addicted to the habit, and they continually prescribe opium for insufficient causes or without any real excuse."[11]

Wright traveled the world on behalf of the United States, and that gave him some much-needed perspective. Wright was the first to understand that the real problem was opioid overprescription rather than smoking. "The history of the opium fight forms a queer illustration of our National blindness to our own faults," he told the *Times*, "and emphasizes our National tendency to see with an amazing clarity, the sins of others, while remaining blind to our own viciousness."[12]

Wright's efforts led to national legislation greatly restricting the import and use of opiates and successfully curbed the prescription of opiates by physicians. The almost total cessation of opioid prescription by doctors lasted for most of the twentieth century. This change in medical practice occurred in parallel with a cascade of drug criminalization culminating in President Richard Nixon's establishing the Drug Enforcement Agency and declaring a "war on drugs." According to insiders, however, this war was less about reducing addiction than it was about controlling racial minorities and political opponents. "The Nixon campaign in 1968, and the Nixon White House after that, had two enemies: the antiwar left and black people," said John Ehrlichman, a Watergate coconspirator, in an interview for *Harper's Magazine* in 1994. "We knew we couldn't make it illegal to be either against the war or black, but by getting the public to associate the hippies with marijuana and blacks with heroin, and then criminalizing both heavily, we could disrupt those communities. We could arrest their leaders, raid their homes, break up their meetings, and vilify them night after night on the evening news. Did we know we were lying about the drugs? Of course we did."

The reaction to the opioid epidemic of the late nineteenth century had led to both the heavy criminalization of opioids and the cessation of their use in clinical medicine. This swing of the pendulum meant hospitals were full of people on the verge of death, suffering uncontrollably in intractable pain.

A hard correction was needed. But it called for a soft touch.

Cicely Saunders was born into an affluent but unhappy household. Her large house was bedecked with tennis courts and lawns, but her mother had given her up to another woman, only to take her back after growing jealous of the adoptive mother. At school, where she was taller than the rest of the girls, Saunders was always made to feel like an outsider. From early in her life, she dreamed of being a nurse, but her father considered the profession unbecoming of her. She went to Oxford, but when World War II broke out, she enrolled as a nurse in

London. "For the first time in my life, I thought I was really in the right place," she said. However, like so many others prominent in the movement to alleviate affliction, her inspiration was personal, as her life had been interrupted by chronic pain, necessitating back surgery. When the war ended, she went back to Oxford and graduated as a social worker.[13]

During her time as a social worker, she fell into an intense spiritual love with a forty-year-old man who was dying. David Tasma was a Polish Jew who had escaped from the Warsaw ghetto and was entirely alone in his newly adopted country as he lay dying of inoperable cancer in a hospital. "We talked together about somewhere that would be more suitable for him than the very busy surgical ward where he was. We wanted a place not just for better symptom control, but for trying to find out, in a way, who he was."[14]

Before he died, David Tasma told Saunders, "I'll be a window in your home."[15]

These words set Saunders on a path that would forever change how we take care of patients, particularly those at the end of life. She enrolled in medical school at the ripe age of thirty-three. The hospitals that Saunders worked in were brutal, particularly for those with incurable ailments. "Many patients feel deserted by their doctors at the end," she wrote.[16]

In order to better understand the management of pain, after Saunders finished medical school in 1958, she undertook a research fellowship with a charitable facility run by nuns. It housed those dying, mostly of cancer, who were expected to live for less than three months. Only 10 percent of her patients there made it past those three months; the chief complaint of 70 percent of the patients in this facility was pain. Here Saunders developed the concepts that would come to revolutionize how pain is managed in those with incurable illness. As opposed to waiting for someone to ask for pain medications, she advocated giving medications in a scheduled manner to preempt a pain crisis. Her drug of choice was heroin. This approach grew out of a new medical concept Saunders developed: the concept of total pain.

Saunders would often recall the tale of Ms. Hinson to describe what total pain meant. When Saunders asked about her pain, Hinson replied, "Well, doctor, it began in my back but now it seems that all of me is wrong."[17]

For Saunders, the pain that many of her patients at the end of life felt was not just a physical sensation. Rather it "presented as a complex of physical, emotional, social and spiritual elements." "The whole experience for a patient includes anxiety, depression and fear; concern for the family who will become bereaved; and often a need to find some meaning in the situation, some deeper reality in which to trust."[18]

To conquer total pain, Saunders advocated a totalitarian approach. "Constant pain calls for constant control," she wrote in a seminal paper in 1963, "and at this stage that means the regular giving of drugs."[19]

Saunders's approach to managing pain was revolutionary, and her advocacy for powerful opioids was completely antithetical to convention. In response to the maelstrom of opioids that had gripped the United States at the start of the twentieth century, doctors had been exceedingly miserly about giving them, reserving them only for the very anguished. Even when provided with narcotic painkillers, patients either were given such small doses or received them after they were already in such deep distress that the drugs appeared to be ineffective. Saunders flipped that strategy on its head.

"There is no such thing as intractable pain, only intractable doctors," she famously said.[20]

Saunders worked tirelessly to ease other physicians' fear of pre-scribing opioids. "Patients may indeed be physically dependent on the drugs but tolerance and addiction are not problems to us, even with those who stay longest."[21]

Saunders went on to build the world's first hospice, a place where people with a terminal, often very painful condition could live and die in comfort. She became the torchbearer of a new medical specialty, palliative care, whose goal was to improve the quality of life of people living with serious illnesses. She was recognized as a dame by Queen Elizabeth.

Cicely Saunders is also one of my personal heroes, injecting much needed humanity into a field in the thrall of technology. Even as she emphasized the need to clamp down on pain with drugs, she understood that pain was far from a one-dimensional experience. "Mental distress may be perhaps the most intractable pain of all," she wrote, and the answer to it was not always more drugs, because for some "the greatest need is for a listener."[22]

For Saunders, giving opioids was an expression of her devotion to the patient in front of her. Her call for medicine to become more humane, particularly to those facing the end, was heard the world over and shapes practice to this day. When a reporter asked her how she would want to die, she said, "Everybody else says they want to die suddenly, but I say I'd like to die of cancer, because it gives me time to say I'm sorry, and thank you, and goodbye."[23]

She would eventually get her wish. Cicely Saunders developed breast cancer and passed away in 2005 in the very hospice she helped develop.

Saunders's legacy lives on in many ways. She was a devout Catholic and very much opposed the right-to-die movement. She believed that if she could help patients achieve zero pain, then people's yearning for autonomy over how they passed away would extinguish. "Analgesics should be given to prevent pain from occurring," she wrote, "not to control it when it is already present."[24]

The problem she faced was that the effects of many opioids like morphine and heroin seemed to wear off quite quickly, with people's pain often recurring even more violently than before. She corresponded with Napp Pharmaceuticals, a pharmaceutical company that had just been bought by an American conglomerate. Her research associate gave them an idea to develop a long-acting form of morphine.

This drug, a pill that only needed to be given twice a day, would go on to be called MS Contin.[25]

It is important to remember that Saunders's approach to care focused specifically on managing the anguish of those who had only days or months to live. While the concept of total pain is useful in thinking about pain more generally, Saunders never made claims about how to treat the total pain of people who weren't dying. Yet, for pharmaceutical companies like Napp Pharmaceuticals, there was simply not enough money in hospice care. They wanted to expand their clientele to a much larger and more lucrative population: everyone living with chronic pain. Their goal was to take Saunders's approach toward total pain experienced by people at the end of life and apply it to people in every phase of life, treating routine medical conditions as if they were terminal illnesses.

MS Contin was a successful drug, but for its manufacturers there was one drawback: its name. Given that the M stood for morphine and many doctors only ever gave morphine to those who were actively dying, there was resistance among clinicians to using this drug with patients who were not at death's door. So Napp repurposed oxycodone, an older drug that worked very similarly to morphine, into a long-acting formulation called OxyContin.

Napp Pharmaceuticals was owned by Purdue Pharma.

Cicely Saunders began a movement in medicine forged with love, empathy, and kindness for the dying. Purdue Pharma, the company owned by members of the Sackler family, hijacked this movement into a business strategy.

The ethos that Saunders developed to manage intractable, terminal pain was copied word by word to develop protocols for how pain was managed for everyone. Saunders's goal was to treat patients as people; pharma's was to treat them as consumers, blind to the risks of their overconsumption. Yet pharmaceutical companies were able to cloak their true intentions with the saintly halo of Saunders's humanity.

In their drive not just to change how we treat pain but to mutate the entire pharmaceutical industry into its own image, Purdue Pharma got plenty of help, most of all from physicians. While many doctors were motivated to fulfill their core duty—to relieve suffering—many others simply became drug mules in white coats.

When my sister Gohar finished her pediatrics training in Dallas and went to work in a rural hospital in Ohio, one of the most common conditions she treated in newborns was opioid withdrawal. These babies were born to mothers actively abusing narcotics. In one memorable case, she treated a baby who began withdrawing on just its second day of life from krokodil, a homemade opioid that usually includes crushed codeine and household supplies like paint thinner and gasoline. Krokodil, from the Russian word for crocodile, is often called a flesh-eating drug because it's so acidic. People who use it more often die from overwhelming infections than from overdoses. "The mother was injecting into her legs," she told me. "She had ulcers on her legs oozing pus."

"We couldn't allow her into the nursery because of her active infection," Gogo told me. Her voice was somber.

Opioid withdrawal has become devastatingly prevalent among American newborns. Almost one in every hundred babies born in the United States experience it, with almost thirty-two thousand infants suffering from opioid withdrawal after birth in 2016. This occurs even more commonly among children born in rural areas. In West Virginia, for example, one in twenty babies experience what is also called neonatal abstinence syndrome. Symptoms can range from tremors, fast breathing, sneezing, diarrhea, excessive crying, and irritability to seizures and death. If babies demonstrate symptoms of severe withdrawal, which many born to mothers using heroin or fentanyl do, the treatment is unfortunately the very drugs that made those children so ill in the first place.[26]

The mother of the child withdrawing from krokodil was admitted to the hospital for her leg infections, but she left abruptly against the advice of her doctors. After a few days, the nursery received a call from one of the mother's family members. She had died of an overdose.

"Some kids just have bad luck from the get-go," Gogo said. In many parts of the country, including the one Gogo works in, the opioid epidemic has overwhelmed the social support system. "We don't have enough foster families to take in babies."

There is a direct link between stories like Gogo's and the actions of members of the Sackler family. The patriarchs of the family were three brothers: Arthur, Mortimer, and Raymond. They were born in Brooklyn to European immigrants who ran a corner grocery store. All three trained as psychiatrists.[27]

Although Arthur, the oldest of the three, died in 1987, before OxyContin was developed, he arguably created the blueprint for its success.

At the dawn of the twentieth century, the pharmaceutical industry was helmed by snake oil salesmen. The market was dominated by manufacturers of "patent medicines" that contained undisclosed ingredients and were marketed as cures for all ailments. One such medication was "Neuralgine," referred to as "the Great Pain Cure" in magazine advertisements, which promised immediate relief for almost every form of pain. Ads for patent medicines accounted for almost half of newspapers' entire advertising revenue.[28]

In response, chemists like Charles Pfizer and Eli Lilly founded small apothecaries with the purported goal of increasing transparency about ingredients and improving the consistency of products. Hundreds of such companies popped up, but most were small and struggled until the outbreak of World War II. The US government desperately needed to boost the production of the first truly blockbuster drug—penicillin—and awarded contracts to fifteen American firms to produce it in large quantities. This massive funding boost transformed the industry into an oligopoly in which fifteen firms now reaped 90 percent of all profits. The pharmaceutical industry, which hadn't existed as such only two decades before, became the most profitable industry in the United States after World War II, retaining that position for the rest of the century.[29]

As the federal government became more involved in this newly dominant field, it attempted to dissuade people from self-medicating, with many drugs now requiring a prescription from a physician, opioids being some of the first. As the industry could no longer sell many of its products directly to patients, it had to focus on selling to doctors, which meant the approach to marketing would also have to change. Cashing in on this shift would require someone with a deep understanding of both medicine and marketing and of both doctors and patients.

Even when Arthur Sackler was in medical school at New York University, he moonlighted as a copy editor in an advertising firm. Just a couple of years after receiving his degree, in 1942 he joined the four-person William Douglas McAdams advertising agency, which specialized in the still-nascent field of medical advertising. By 1949, Sackler had made enough money to own the company.

He possessed a preternatural understanding of both the human mind and the modern financial ecosystem. While doctors prided themselves on their rationality, it turns out they were just as susceptible to advertising as anyone else. Arthur Sackler used medical journals as a vehicle for marketing his drugs to doctors. This was a smart strategy. Ads in journals were exempted from regulation by the Federal Trade Commission because doctors were thought to be able to objectively evaluate the pros and cons of medical therapies. In fact, the return on investment for every dollar spent on advertising in medical journals is higher than for any other form of medical advertising. In 1958, the industry paid for almost four billion pages of advertisements in medical journals, and medical marketers made twenty million calls to doctors and pharmacists.[30]

Among Sackler's early successes was the marketing of Valium, which he successfully branded as an antianxiety medication and turned into the first billion-dollar drug in the history of pharmaceuticals. Valium was one of the first psychotropic drugs that you acquired not in a back alley but at a pharmacy. The true success of pharmaceutical marketing was transforming the patient into a consumer, making pill popping and overprescription a way of life.

In 1932 Aldous Huxley published his novel *Brave New World*, which describes a dystopian future in which autocrats modulate the entire range of human emotions and feelings with a designer drug called soma. Soma isn't a medication for a specific condition or for someone sick. Yet there isn't a condition soma can't ameliorate or a person it can't help feel better. Soma is an instrument to change one's reality, to give one "a holiday from the facts," Huxley writes. "In the past you could only accomplish these things by making a great effort and after years of hard moral training. Now, you swallow two or three half-gramme tablets, and there you are."

Huxley's speculative fiction couldn't have better predicted the years to come. World War II had seen the development of antibiotics, used primarily to treat infections or prevent injuries and wounds from getting infected. But as the war ended and waves of American troops came home, they returned not just with mutilated limbs but with deep mental scars.

Psychiatric institutions quickly filled up with men and women with profound maladies like schizophrenia. One such facility was Creedmoor Psychiatric Center in Queens, New York, where Arthur Sackler trained. He subsequently directed the Creedmoor Institute for Psychobiological Studies from 1949 to 1954, where all three brothers worked.

At Creedmoor, as was usual then, patients were shackled and sedated, castrated and lobotomized, in efforts to tame their wayward instincts. A drug for the severely insane would be helpful but would always have limited reach if it couldn't be marketed to the public at large.

As part of his work selling antibiotics for Pfizer, Sackler pushed for a Pfizer marketing booklet to accompany every issue of the *Journal of the American Medical Association*. He also launched *Medical Tribune*, a free newspaper for doctors, reaching hundreds of thousands and helping to advance his pro-prescription agenda. This publication was notorious for disseminating misinformation that generally favored pharma's bottom line. This included fearmongering about generic drugs, with one false story, headlined "Schizophrenics 'Wild' on Weak Generic," intended to scare doctors about how "all hell broke loose" at a facility when patients were switched to a cheaper generic antipsychotic drug. An advertisement for the Pfizer drug Diabinese, marketed by Sackler's agency, proclaimed an "almost complete absence of side effects," even though 27 percent of patients in fact did develop serious side effects.[31]

Arthur Sackler's business was mired in scandal. One company he owned, MD Publications, paid hundreds of thousands of dollars to Henry Welch, the head of the antibiotics division at the Food and Drug Administration (FDA), in exchange for his support for certain drugs. The resulting outcry led to Welch's forced resignation.[32]

The "education" pharma promoted served an even more insidious purpose: it trained physicians to profile the people coming to their clinics or hospitals, nudging them to make at-a-glance diagnoses. And among the recurring images in advertisements for Valium were stressed-out women, often depicted as "neurotic singles, tired middle class mothers, exhausted business women, and irritable menopausal women." After they graduated, physicians received little high-quality new research and often relied on advertisements and marketing promotions to update their practice. The marketing campaign for Valium taught physicians to treat not people but stereotypes. To further refine marketing targeting physicians, Sackler founded IMS Health, a prescription data company that could identify what one Pfizer memo described as "easy prey."[33]

Sackler now influenced every aspect of the healthcare ecosystem. And his marketing campaign for Valium was proof of his dominance. Valium was the best-selling prescription medication between 1968 and 1981, with more than two billion tablets sold in 1978 alone. Moreover, it became a cultural staple. The Rolling Stones crooned about "mother's little helper." Popping pills for frayed nerves became a national habit—but at a huge cost. The most famous celebrities of their time, Elizabeth Taylor and Elvis Presley, were both addicted to Valium. To this day, more than twelve million Americans report abusing diazepam (Valium) every year.[34]

Sackler "invented the wheel of pharmaceutical advertising," and this wheel has rolled over everything in its path, including, ultimately, people, by redefining the patient as a consumer.[35]

"Valium was, significantly, one of the first psychoactive drugs to be used on a large scale on people who were basically fine," wrote Robin Marantz Henig for the *New York Times* in 2012. "It made people wonder what 'normal' really was."

Pharmaceutical advertisers rode the wave of unlimited wants in modern industrialized societies built around marketing strategies that constantly push people to mull over what they lack. Rather than discover new drugs, marketing helped people discover new yearnings. It wasn't enough to be rid of heart disease or cancer. In the utopian dream right out of the pages of *Brave New World*, the drug industry now made us want to be free of sadness, grief, inattention, boredom, stress, and, of course, pain.[36]

The United States and New Zealand are today the only countries that allow drugmakers to directly advertise to consumers. Their sales pitches have been so potent that over time Americans' responsiveness even to placebos has increased, while in other countries it remains the same. This sustained marketing blitz means that when the average American takes a pill, *any pill*, their expectation of relief from pain is greater than that of someone from another country.[37]

More than anything else, pharmaceutical marketing has narrowed the scope of what is deemed an acceptable state. When does grief become major depressive disorder? When does apprehension become generalized anxiety disorder? When does inattentiveness become attention deficit/hyperactivity disorder? And when does an ache become chronic pain?

While Sackler managed to avoid scrutiny through much of his life, he did catch the eye of a senator who had built his career prosecuting a business with eerie similarities—the Mafia. Estes Kefauver was a prominent Democratic senator from Tennessee who became famous as the chairman of the Senate Special Committee to Investigate Organized Crime in Interstate Commerce, bringing in mobsters like Frank Costello to testify on national television. The Democratic Party's candidate for president in 1952, he was one of only three southern senators to oppose efforts to racially resegregate schools.

After focusing on the Mafia, Kefauver turned his lens on the pharmaceutical industry, understanding the paradox of this burgeoning field. "The drug industry is unusual in that he who buys [the patient] does not order [the doctor], and he who orders does not buy." The hearings continued for almost two years, and while they initially concentrated on drug costs, they evolved to focus on something few knew at that time: there was no mechanism to know if drugs reaching the market even worked or not. At that time, the FDA had no authority to require a drug's proof of effectiveness. And while the FDA supported an expansion of its authority, a big opponent of this push was, sadly, the American Medical Association, the largest physician lobbying organization, which took that stance in part because Kefauver advocated for restrictions on medical advertisements, which to date remain a robust revenue stream for medical journals.

Despite this opposition, the Kefauver Harris Amendment to the Federal Food, Drug, and Cosmetic Act was passed and signed by John F. Kennedy in 1962. The amendment introduced the proof-of-efficacy requirement for drugs to be approved. Notably, the amendment also placed restrictions on claims that drugmakers could make in their marketing.

It was during these hearings that Kefauver crossed paths with Arthur Sackler. When Sackler was called to testify in front of the Senate, Kefauver couldn't pin him down, even though a memo prepared by one of Kefauver's staffers perfectly summarized the vast, byzantine network that Sackler had put together:

> The Sackler empire is a completely integrated operation in that it can (1) devise a new drug in its drug development enterprise, (2) have the drug clinically tested and secure favorable reports on the drug from the various hospitals with which they have connections, (3) conceive the advertising approach and prepare the actual advertising copy with which to promote the drug, have the clinical articles as well as advertising copy published in their own medical journals, [and] (4) prepare and plant articles in newspapers and magazines.[38]

This memo should have been a wake-up alarm, but it went unheard. The numbered points would serve as the four pillars of the strategy used by Purdue Pharma and other pharmaceutical manufacturers to hook the world on their products.

In late 2016, while I was training at Duke University Hospital, legendary NPR broadcaster Diane Rehm invited me to come on her show to discuss physician-assisted death, something I supported. To debate the opposing position, she invited a much more senior physician. Early in the interview, he began talking down to me. "I wished that I was a faculty member at Duke at the time, rather than a faculty member of Dartmouth, where I'm an emeritus professor," he told me condescendingly, suggesting that I was speaking from a place of inexperience. My stance in favor of physician-assisted death came from "common concerns that doctors in training have that we regularly, you know, correct."

We ended up talking about morphine, and he said categorically, "Used correctly, unless it's really intended to end life, morphine and other pain medications are entirely safe and can be used to alleviate suffering." When I went on to say that escalating doses of morphine in a drip form could cause people to eventually stop breathing, he cut me off.

"That's just not true," he interjected. "That's simply not true."

In that very year, when that physician told me emphatically that morphine was "entirely safe," 42,249 Americans died of opioid overdoses.[39]

He went on to say with godlike certitude, "We can alleviate all physical distress."

The medical profession had come full circle to where we were at the end of the nineteenth century. Somehow we had again arrived at the belief that opioids were a complete cure for pain. This entirely untrue and indefensible position was held not just by some corrupt doctors bankrolled by pharmaceutical companies but by many compassionate and well-meaning ones. Many of them were my teachers and mentors.

Just before Cicely Saunders's revolution, physicians wrongly hesitated to prescribe opioids to people who were on the verge of death. Contrast this to 2016, when American doctors wrote 214 million prescriptions for opioids, translating to sixty-seven prescriptions per one hundred Americans, only a small fraction of whom had a terminal condition. In several states, more prescriptions were written than there were people. In 2016 alone, 11.5 million Americans reported misusing prescription opioids.[40]

While the vast majority of these prescriptions were for patients with chronic pain who didn't have cancer, decades of research now conclusively suggest that opiates provide little to no benefit for chronic noncancer pain. One recent randomized trial showed pain intensity was worse in US veterans using opioids for moderate to severe chronic back and joint pain than in those using medications like acetaminophen and ibuprofen. There is evidence that rather than improving chronic pain, opioids could actually make it worse.[41]

Why is this? Studies have shown that opioids reduce patients' pain thresholds, making them increasingly sensitive to any stimulus. They can also cause opioid-induced hyperalgesia, a condition in which people feel more and more pain as they take higher and higher doses of opioids. So while opioids do help many patients suffering acute pain from injuries, surgeries, or extreme conditions like cancer, their indiscriminate use to treat chronic pain—backaches, arthritic knees, and the like—is arguably the worst medical mistake of all time.[42]

When I was in my primary care clinic, a new patient saw me one afternoon. She split her time between Boston during the vibrant spring and summer and Florida during the frigid winters. She came to see me because she wanted to have a primary care physician who knew her well while she was in Boston. We chatted about when she'd had her last pap smear and whether she'd been tested for high cholesterol or diabetes. She seemed to be in great health. As the visit was ending, she told me she needed a refill for her oxycodone.

"Why do you take Oxy?" I asked.

"Oh, it's for my back."

I asked for a prescription. She produced a handwritten scribble on a piece of paper that she had gotten from a pain clinic in Florida.

"I've been going there for more than ten years," she told me.

I told her I would be happy to renew her prescription but wanted to first get in touch with her clinic in Florida so I could confirm why she had been getting the oxycodone.

When I called the phone number, the staff told me they knew the patient well. I asked them to share all the records they had of her ten years as their patient. I parked myself by the fax machine as the first sheet of paper started to scrunch its way out. It had some basic information: her name and date of birth, a mention of back pain, and a prescription for oxycodone. I waited for the entire stack of papers to come out. I could only imagine what ten years of medical records might look like; I wouldn't have been surprised if there were hundreds of pages. Yet, for this patient, this was the only page that ever came.

I called the clinic to say there had been an error. What they told me in response shocked me: that single page was her entire medical record.

It was clear to me in that moment what had happened. I had reached into one of the dirtiest corners of our rotten healthcare system: the pill mill.

Medicine can only exist as a profession if it is enmeshed in a moral code. For a person to walk into a clinic and allow someone they have never met before to give them a drug they know nothing about, they must believe that the doctor or nurse who gives them that drug is driven by no impulse other than to seek what is in the patient's best interests. That trust is what makes the system work.

Medicine doesn't just attract people who want to help others but rigorously trains them in shaping their instincts—to never be too exhausted, overwhelmed, frustrated, or tired to make the most ethical and informed decision possible.

What would have to happen, then, for a doctor to set up a clinic that essentially acts as a sanitized front for a drug-dealing operation, where patients by the thousands are given tranquilizers, stimulants, and narcotics by the bagful? Who is responsible for the misery they bring to people's lives?

Pill mills are just one manifestation of the fracturing of medicine's goodness. This fracturing was far from passive. The traditional moral edifice of medical practice had to be actively demolished before it could be replaced with the medical-industrial complex. Over the course of the twentieth century, pharmaceutical companies have not just created a culture of medical overconsumption but effectively changed all healthcare into one giant pill mill.

Ivan Illich, the firebrand social critic who wrote *Medical Nemesis*, was written off as an alarmist. Yet he couldn't have been more prescient. As new tranquilizers such as Valium were developed and age-old drugs such as opium, known to be both dangerous and addictive, were redesigned as even more potent formulations, such as heroin and morphine, pharmaceutical companies faced a problem: How could they convince physicians that everything they'd learned about the dangers of opioids wasn't true of these new products?

To accomplish this, the medical-industrial complex would have to rewrite everything we had learned about pain.

7.

CROWN OF THORNS

The Hijacking of Modern Medicine

In the face of pain there are no
  heroes. ·
                              —George Orwell

MATT SUFFERED CHRONIC DAILY HEADACHES. His doctor had prescribed him almost every narcotic available to treat them—morphine, meperidine, codeine, oxycodone—but his favorite was hydromorphone, a powerful opioid that also went by its brand name, Dilaudid. "I started off just snorting the D as I had been with the morphine, but eventually my tolerance started to grow so I knew I needed something more because snorting dilaudid is only 50% as effective as shooting," he wrote in a post for the website Erowid.org. "So I went to the clinic for an app[ointment] and while I was waiting for the doc I stole some syringes that were sitting in the cupboard, went home, locked myself in the bathroom, and tried whacking the dilaudid."

Matt crushed a few tablets and injected them into a vein in his left arm. "The feeling of slamming dilaudid is amazing, it starts off as a slight warmth and creeps up to my brain until finally my brain goes off with a gentle explosion of warmth that fills my entire body from head to toe, into my fingertips, it envelops my whole body," he wrote. "I was in heaven."

Matt's mom eventually found his syringes and confiscated most of them. But he still had a few left, which grew increasingly blunt with his frequent use. "They'd bruise the living shyte out of my veins, and tear my skin as I stuck them in."

As the effects of even the Dilaudid started to wear thin, he went back to his doctor, who prescribed him a long-acting form of the drug. His arm started to turn yellow from all the pills he had been injecting.

Matt kept abusing the Dilaudid until one day he had a complete psychotic breakdown: "I slit my wrist twice that day, tried to stab my brother and mom, and they called the cops. 2 paramedics and 4 cops came and dragged me to the hospital where I was screaming things like 'I HOPE YOU DIE' and 'JUST ONE MORE FUCKING HIT' they heavily sedated me on lorazepam and forced me to stay in the psych ward for about 4 days, and prescribed me a lowering dose of dilaudid which they watched me take."

Matt signed off his post with a quote from Gus Van Sant's critically acclaimed film *Drugstore Cowboy* (1989): "Most people don't know how they're going to feel from one minute to the next. But a dope fiend has a pretty good idea. All you gotta do is look at the labels on the little bottles."

Matt is but one of the millions upon millions of Americans prescribed opioids, often inappropriately, often recklessly, often dangerously, leaving us positively awash in these drugs. After my wife gave birth to our daughter, she was given a thirty-day supply of oxycodone. She never filled the prescription because she didn't need to. We never let opioids enter our home. But for many others, opioids are always at hand.

And for many people struggling with opioid addiction, the story is a recurring one: An individual goes to the emergency room or doctor's clinic with some type of pain or injury and is given a prescription for opioids. Over time their tolerance increases; the same dose just doesn't cut it. In between doses, they begin to have symptoms of intense physical and psychological withdrawal. Opioids actually reduce the user's pain tolerance, so the patient's chronic pain worsens. To avoid

THE SONG OF OUR SCARS

withdrawal, to reduce how much pain they feel, and to overcome the tolerance they develop, the patient may take more opioids, only to eventually run out.

At this point, regardless of what happens next, the patient with chronic pain is left in far worse shape than before. The patient's doctor now faces two bad choices: either increase how much opioid is prescribed or make no changes. Regardless of the physician's decision, a downward spiral awaits. And at that point some might turn to a different source for their opioids. Any change in an opioid prescription—be it cessation or an increase in the dose—raises the chance that the user will transition to seeking opioids from a dealer rather than a doctor.[1]

This journey is not just one that many individuals take; it also reflects how the broader opioid epidemic has evolved. As doses escalate due to rising tolerance, worse becomes the withdrawal and stronger the urge to take a larger dose, bringing death only a whisker away. And as people require increasingly hefty doses of prescription opioids, these pills become a gateway to far more dangerous drugs like heroin and fentanyl. As a result America's streets have become overrun with cheaper yet more potent opioids, spiking the death rate from narcotics.[2]

Injectable drug use, spurred by overprescription of opioids, has become such a scourge in the United States that it has led to multiple medical crises. Both HIV and hepatitis C have proliferated, as have life-threatening infections of the heart.[3]

The reason opioids have become such a recurring nightmare in human society is that they are really good at what they do. In the acute setting, when a patient has a broken leg, a hernia threatening to burst, or a cancer spreading into their bones, opioids might well be the most effective therapy.

Yet, as incredible as opioids are at treating acute pain, they can be woefully inadequate and downright lethal for many with chronic pain.

When the human body detects an opioid swimming through its bloodstream, it doesn't see a stranger. Opioids, as we are learning, not only play an integral role in how our bodies overcome pain but might well be the essence of our humanity.

It might be a cliché to say this, but everyone hurts in their own way. Two people may suffer the same injury, but one might feel a lot of pain, and the other might not. One long-distance runner could keep running despite discomfort that would cripple another.

Our innate ability to endure pain is so immense that it may well be considered a superpower. And it comes from our body's ability to produce its very own supply of opioids. Within all of us is a universe of opioids and receptors that mediate not just all that hurts but all that pleases. The spectrum of pain and pleasure, the push and pull of desire and disgust, the very question that vexed Socrates, is primarily a measure of the balance of endogenous opioids in our bodies.

Balance in the endogenous opioid system is the key to our feeling of normality, of peaceful stillness in the moment. When we take opioid pills or injections, that balance is changed, and for some, the change is permanent.

Opioid receptors are widely distributed in the human nervous system and, indeed, have likely been present in vertebrates for at least 450 million years. When Patrick Wall and Ron Melzack conceived their gate theory of pain in 1965, while their overall hypothesis was sound, they knew little about the actual components of the pathways they proposed that pain traveled on. We know now that endogenous opioids are the keys to the gates that pain has to pass through to be felt.[4]

The first and perhaps most important parts of the body's endogenous opioid system are the receptors that opioids like morphine bind to. Initially called the Mμ receptors, they were recently rechristened as MOP. The primary internal opioid that acts on MOP is called beta-endorphin.

When stimulated in the midbrain, MOP receptors send signals down to the spinal cord, slamming the door on painful nociceptive signals attempting to ascend the nervous system toward the brain. And this is how opioids essentially work: they overwhelm these gates, shutting them tight, preventing nociception from transforming into pain by reaching the brain.

Beta-endorphin is more than our very own painkiller. It is also the key facilitator of the feeling one gets when eating a delicious meal or making love or when holding one's child really close. Endorphins are the chunk of cheese at the end of the maze, the summit at the end of the trek, the trophy at the end of the tournament.

The endogenous opioid system is also closely linked to our stress response system—one might well be the yin to the other's yang. Understanding this relationship is important to achieving a greater understanding of both chronic pain and the effects of opioid use on how we feel.

While in everyday parlance, stress has a decidedly negative connotation, our body's ability to respond to crises is critical to its survival. When we encounter the sight or even the smell of a predator, hormones like cortisol and norepinephrine mobilize our immune system, raise our blood pressure and heart rate, and increase our attentiveness. The same intricately evolved response helps us fight off a serious infection. Yet a persistent stress response is linked to depression and posttraumatic stress disorder, as well as heart disease, obesity, and irritable bowel syndrome, to name just a few. Therefore, after a stressor subsides, we need to douse the flames of the stress response, and that's where endogenous opioids come into play.

When we feel stressed, our body produces a molecule that is split into ACTH, a stress hormone, and beta-endorphin, our chief endogenous opioid. So even as the stress response is revving up, our inner opioids modulate how extreme that response will be and how quickly we will recover from it. The influx of endogenous opioids during stress is also why we feel less pain during a calamity, allowing a wounded soldier to escape the battlefront. In one experiment, novice parachutists showed decreased pain sensitivity because of the anxiety they felt, an effect that went away when they were given a drug that blocked the effects of opioids. Our inner opioids serve to take the edge off, to satiate, to unplug.[5]

Yet just as modern life is characterized by chronic rather than acute pain, the same is true of stress. Instead of the occasional predator, today's stressors are an incessant barrage of work emails, the escalating demands of domestic life borne increasingly by the individual rather than shared by the community, and rising anxiety wrought by a world that is increasingly digitally connected but interpersonally fractured. This state of constant stress causes us to continually produce endogenous opioids, both rendering their antistress properties less potent and inducing opioid dependence, which increases our sensitivity to everyday nags and nicks. An ache that might never even be registered by the person at peace might crush the person under duress.[6]

Even as chronic stress can predispose people to developing chronic pain, chronic opioid use can make the problem worse. This might be surprising because, given how central endogenous opioids are to our own ability to kill pain, one would surmise that prescription opioids would be a panacea for chronic pain. Yet a recent review of 162 studies by the US Agency for Healthcare Research and Quality showed that for patients with chronic pain, opioids were barely better than placebo at alleviating pain or restoring physical function and no different at improving mental well-being. And they were no better than nonopioid pain medications like acetaminophen (Tylenol) or ibuprofen. This lack of benefit was accompanied by significant harm and a plethora of side effects.[7]

The reason opioid medications don't work for chronic pain is that they are simply too blunt and too powerful, rocking the delicate balance of the body's natural pain-regulation systems. Even a relatively small dose of a drug like morphine or oxycodone dwarfs the amount of beta-endorphin our body can produce by itself. To our body, a prescription dose of opioids is like a snowstorm in Texas, overwhelming the infrastructure. And so the body goes to work to restore equilibrium by reducing the number of opioid receptors, which dulls the impact of the drugs. This is why, over time, we develop a tolerance for opioids and need a higher dose to achieve the same effect.[8]

Yet this change has another more insidious effect: chronic use of opioids leads our body to produce less of its own endogenous opioids and other related chemicals.[9]

This makes people who take opioids more likely to experience exaggerated stress responses that are not counterbalanced by endogenous opioid production, which also means it takes them longer to recover from such episodes. The link between depression and chronic pain is also mutually reinforcing. People with depression are more likely to produce lower amounts of endogenous opioids, helping explain why opioid drugs often help depressed people feel better; however, by causing a reduction in endogenous opioid production, chronic opioid use further suppresses their ability to feel better on their own, which further deepens their depression.[10]

Looking closer into the science of our inner opioids reveals that what manifests in our bodies and what is expressed in our soul are more connected than we could have ever imagined. Endogenous opioids are the reason we get a sense of social connection and belonging. Prescription opioids monopolize the feeling our strongest life experiences provide us, leaving users unable to experience them on their own. A mother feels an indescribable warmth when she holds her child and a deep sense of loss when her newborn is snatched away from her, but morphine stamps that instinct down. That warm feeling experienced when reading a loving message from a friend or family member is also a function of opioids, and we lose it when we take narcotic painkillers.[11]

When our bodies are first exposed to opioids, relief from all suffering begins to circuit through our veins. Many studies have shown the effect to be similar to that of orgasm. Opioids make pain better, food yummier, relationships deeper, and the soul happier. As they wear off and withdrawal kicks in, not only is the pain worse, but food turns bland, relationships become an afterthought, and the spirit withers. Everyday traumas become catastrophes as our bodies lie helpless, deprived of our natural means of respite. Opioid withdrawal doesn't just make us more sensitive to pain than we ever were before; it also makes us more sensitive to negative emotions, a phenomenon called hyperkatifeia. The only thing that will halt this descent is another dose.[12]

Put it all together, and it becomes apparent why chronic stress leads to chronic pain, how our state of mind can overwhelm the state of our bodies. As we experience bout after bout of emotional upheaval, not only are we continually manufacturing stress hormones, but we are also trying desperately to maintain balance by overproducing endogenous opioids. The persistent elevation of stress hormones leads to the persistent production of opioids, which has a similar effect to when we persistently use opioid painkillers: the effect of endogenous opioids diminishes, while our need for external opioids increases, and our ability to stave off daily aches and pains using our own defense mechanisms vanishes. Therefore, a life of relative physical comfort but emotional distress can become overwhelmed with both emotional and physical distress. Individuals who experience recurrent psychological trauma, such as those living with posttraumatic stress disorder, are more likely to use opioids for pain relief and exhibit a higher tolerance for opioids when they do. And as they use even more opioids, their stress hormones surge to counteract the opioids, leaving them feeling even more strained than before.[13]

When people become trapped in the clutches of chronic pain and chronic stress, their lives become engulfed in an eternal, inextinguishable fire—a fire as ferocious as the one that drives the hunger for profit that fuels many modern pharmaceutical giants. Yet it would be shortsighted to just blame drugmakers without identifying the many forces that appear to be driven by a searing thirst for revenue. Companies like Purdue Pharma are the apex predators best adapted to the incentives at the heart of modern healthcare. Though Purdue has declared bankruptcy in the aftermath of numerous lawsuits, the system that continues to mistreat and misuse the person in extremis continues to thrive.

The success of Valium, a tranquilizer, proved to the pharmaceutical industry that the path to expanding its empire lay not merely in developing therapies for the ill but in expanding the very definition of illness. To do so pharma didn't just need marketing; it needed a movement. Instead of building one from scratch, it commandeered one that was already taking off.

Cicely Saunders had shed light on the excruciating agony regularly faced by those dying of cancer. She helped initiate a movement that hoped to ensure comfort and dignity for those with terminal illnesses. While Saunders and many other doctors and nurses saw this movement toward a more patient-centered practice as an opportunity to infuse much-needed compassion into medical care, others saw something completely different: a financial windfall. Repackaging existing drugs to treat those suffering at the end of life was lucrative, but not as lucrative as if the scope of such aggressive treatment could be expanded to a larger group of people who had longer to live, which meant that they would remain customers for a longer period of time.

The pharmaceutical industry realized that by convincing doctors to treat people with musculoskeletal pain the way Saunders had advocated treating the terminally ill, it could greatly expand its market. So it aligned itself, in appearance anyway, with Saunders, one of the most trusted voices in medicine.

Convincing doctors and nurses to more aggressively stamp out pain was easy, for it appealed to the well of empathy that exists in most clinicians. Most join the healing vocation to help people overcome suffering in any form. However, the majority of clinical training is strictly quantitative: learning the pathways of biochemicals, the routes of nerves and tendons, the vagaries of medical statistics. We are taught to read research papers as if they were inscrutable religious tomes. Within every clinician, thus, is a voice always whispering, "Show me the data."

This was a problem for pharmaceutical companies: they had no research showing that opioids were either safe or effective for patients with chronic pain. So they had to lie and to find doctors who would help.

In 1980, two researchers from Boston Medical Center wrote a five-sentence letter in the *New England Journal of Medicine*. In an analysis of almost twelve thousand hospitalized patients who had received a narcotic, they found only four cases of addiction. "We conclude that despite widespread use of narcotic drugs in hospitals," they wrote, "the development of addiction is rare in medical patients with no history of addiction."[14]

The problem with this strong statement was that the letter was not designed to be able to make that assertion. The data collected weren't even designed to look for addiction. Since then, one of the authors of this sloppy letter, which he now believes "wasn't of value to health and medicine in and of itself," has claimed that he wishes he hadn't published it. The journal also published a disclaimer in 2017 that sits right above this letter: "For reasons of public health, readers should be aware that this letter has been 'heavily and uncritically cited' as evidence that addiction is rare with opioid therapy."[15]

Such repudiations, though, came almost four decades too late. The letter, which incorrectly suggested that less than 1 percent of opioid users get addicted, became the lynchpin of decades of frothy-mouthed marketing. Other physicians too began to report similar poorly performed research. One such paper was written by New York–based physicians Russell Portenoy and Kathleen Foley.[16]

In 1986 Portenoy and Foley published their own experience with thirty-eight patients without cancer who were using opioids. They reported noting no single episode of an adverse effect and concluded, "Opioid maintenance therapy initiated for the treatment of chronic non-malignant pain can be safely and often effectively continued for long periods of time." They cited their own previous work to state, "Neither tolerance nor physical dependence necessarily pose difficulties in management." They made an analogy with alcohol, writing that "most people who drink do not become alcoholic."[17]

For their work Portenoy and Foley were celebrated not only by academia but also by pharmaceutical companies. Portenoy was paid heavily by, among others, Purdue Pharma, proselytizing about the benefits of opioids. *Time Magazine* dubbed him the "King of Pain." Among other achievements, he helped author clinical guidelines that Purdue Pharma essentially used as a sales guide. Portenoy recently switched sides and is testifying against pharmaceutical companies. He politely declined my interview request, citing "legal matters still outstanding."

Kathleen Foley has had a long and celebrated career and garnered numerous awards, including a humanitarian award from the American Cancer Society. One of her greatest supporters, though, was Purdue Pharma. Over the years, she published several papers funded by the company and gave numerous talks that supported the industry's position to expand the use of opioids. Her voice reached all the way to the top of the Purdue pyramid. In 2001, when newspapers first began to report the dangers of OxyContin, she wrote in an email to Richard Sackler,

> I'm thinking of an alternative strategy of bringing together all of the members of the pharmaceutical industry, who have analgesic drugs out there and try to come together as a sort of cohesive voice recognizing that your particular drug has been recently identified in the newspapers as a drug issue. I think that there is a tightrope that you need to walk, because you are a drug company and it would be much better if the advocacy came from outside of the drug company and even better without much in the way of support from you…and strategize around the way to play the media issues.[18]

Purdue and other opioid manufacturers took the advice to heart, pumping millions of dollars into patient advocacy organizations that supported their positions. A congressional report found that five opioid manufacturers paid patient advocacy groups and affiliated doctors more than $10 million between 2012 and 2017, with Purdue paying almost half of this. The report concluded, "These groups have issued guidelines and policies minimizing the risk of opioid addiction and promoting opioids for chronic pain, lobbied to change laws directed at curbing opioid use, and argued against accountability for physicians and industry executives responsible for overprescription and misbranding."[19]

Through a $1.5 million gift, Purdue even funded the Kathleen M. Foley Chair for Pain and Palliative Care at the Center for Practical Bioethics in Kansas City, a center with extensive connections to other drug manufacturers as well.[20]

Kathleen Foley was anointed as a leader of the pain movement and shaped policy worldwide. She served as a past president of the now defunct American Pain Society, an organization that frequently took pro-pharma positions, and helped establish the American Pain Foundation, also now shuttered. A report by US Representatives Katherine Clark and Hal Rogers, titled *Corrupting Influence: Purdue and the World Health Organization*, singles her out as one of the key agents influencing opioid policies globally.[21]

As pressure mounted, pharma's representatives in academia deflected blame for the opioid epidemic away from the companies toward their victims: patients.

Court documents have now revealed that Richard Sackler's key strategy to fend off criticisms against OxyContin abuse was to blame the people who died and not the drug that caused their demise. When an acquaintance wrote in an email, "Abusers die, well that is a choice they made, I doubt a single one didn't know of the risks," Sackler wrote back, "Abusers aren't victims; they are victimizers."[22]

While Richard Sackler expressed such views in private, Foley did so in public. The Food and Drug Administration launched an initiative requiring opioid manufacturers to have a Risk Evaluation and Mitigation Strategy (REMS) prior to approval of new opioids. A REMS can include maintaining a record of patients receiving opioids, which can be used to detect adverse events, and providing education for clinicians about risks and benefits of medications. Foley, however, was a vocal critic. "REMS pits the interests of two communities against each other—cancer patients and drug abusers—and potentially stigmatizes millions [of] patients without necessarily reducing risk diversion of drugs and abuse," she said. On an FDA panel, she echoed Purdue's talking point that diversion of drugs like OxyContin to be used for illicit purposes was a criminal, not a medical, problem.[23]

When I emailed Kathleen Foley for an interview, I never received a response.

Purdue transformed the modern pain movement essentially to sell its product. It was so successful that pain was eventually declared the fifth vital sign at the time of OxyContin's launch in the early 1990s. While many doctors may not have known they were pawns in a greater game, in an unsealed email Richard Sackler revealed that he knew exactly what his money was buying: "Our goal is to bind these [pain] organizations more closely to us than heretofore, but also to align them with our expanded mission and to see that the fate of our product(s) are inextricably bound up with the trajectory of the pain movement."[24]

As the pain movement rollicked on, corrupting everything it touched, it came up against one final bulwark, built to protect patients from unsafe medical interventions. The Food and Drug Administration has a rich and storied history; yet not only did it get ensnared in this manufactured movement, but it helped unleash this national catastrophe.

One afternoon in a sunlit room in Boston, I was attending yet another noon conference with my hands covered in what I would like to hope again was burrito juice. That day, my coresidents and I were learning that everything about the way we were taught to treat pain was wrong.

Traditionally, physicians had treated pain by giving someone medication when they were hurting, just as we'd reach for a pill to help alleviate the aching after twisting our back or pulling our shoulder. But for patients with chronic pain, we were now taught to take a different approach. Instead of waiting for people to have more pain and ask for more opioids, we needed to clamp down on the pain *contin*uously. We needed to give chronic-pain patients drugs with "contin" in their names, such as OxyContin or MS Contin, like clockwork.

In the most prestigious medical school on earth, from the best teachers and physicians, we were unknowingly taught meticulously designed lies.

One such lie was pseudoaddiction. In 1989, two physicians described a single case of a patient who, despite receiving morphine, exhibited behaviors that were often attributed to people addicted to opioids. They defined pseudoaddiction as an "iatrogenic syndrome that mimics the behavioral symptoms of addiction" and argued that the proper treatment for patients demonstrating pseudoaddiction was to give them *more* opioids. The idea was that even when a physician was concerned that their patient was addicted to narcotics, if pseudoaddiction was even a possibility, they should still give them more of the drug they seemed to be dependent on. This concept was never tested formally, but it nonetheless became a standard notion in medical education and was cited hundreds of times in the medical literature. One of the two authors who coined this "influential educational concept commonly used in pain management lectures" was David Haddox, who subsequently joined Purdue Pharma as a senior executive.[25]

Purdue Pharma proved adept at influencing one of the most important public health institutions in the country. The Food and Drug Administration was created to protect Americans from dangerous chemicals in their pantries and their medicine cabinets. A shining beacon in the history of the organization is Frances Oldham Kelsey, who prevented thalidomide, which caused severe birth defects, from being approved for use in America. Through single-minded diligence, Kelsey saved American mothers and their babies from exposure to that poison.

When the FDA approved OxyContin in 1995, there was one line on its label, the leaflet that comes with every prescription, that would fuel the furnace powering OxyContin's success, helping spark the opioid epidemic: "Delayed absorption, as provided by OxyContin tablets, is believed to reduce the abuse liability of a drug." This line essentially allowed OxyContin to be legally advertised as less addictive than any other narcotic.

Prior to the drug's approval, doctors had been wary of opioids, as they correctly worried about their addictive potential. OxyContin was essentially the same compound as the old and cheap opioid oxycodone, which was similar to morphine or heroin. Yet, because OxyContin was longer acting—due to a special coating, its effect occurred more slowly over more hours—the company proclaimed, in the absence of any data or testing, that it was safer and less addictive.

The statement that OxyContin was somehow less prone to misuse was used to overturn decades of hesitancy among doctors and patients about the risks of opioids. Yet the coating could easily be bypassed by crushing the pill. Why would the FDA, which vets every single syllable of a drug's label, allow this line to be printed?

Court documents reveal that on first drafting the OxyContin label in 1994, Purdue did not include the consequential abuse-liability statement. Yet it was present in a tracked-changes version the company sent to the FDA in August 1995.[26]

Former FDA medical official Curtis Wright led the agency's review of OxyContin. He testified that he did not remember who wrote it but said, "The label makes an extremely weak statement about a class of drugs." Two years after leaving the FDA, Wright took a job at Purdue.

The other FDA official who worked on the OxyContin approval, Douglas Kramer, also went on to join Purdue.[27]

When OxyContin's label was revised in 2001, while the abuse-liability sentence was removed, more falsehoods were introduced. OxyContin was approved "for the management of moderate to severe pain when a continuous, around the clock analgesic is needed for an extended period of time," despite there being plenty of evidence that patients experienced significant withdrawal symptoms between doses, creating what one pharmacologist described as "the perfect recipe for addiction." The FDA handed Purdue what former FDA commissioner David Kessler called "a blank check" not once but twice.[28]

As Americans continued to be crushed by opioids, the FDA continued to approve ever more potent painkillers, despite its own experts advising against them. Far from a voice the public could trust, the FDA became an amplifier of the big lie.[29]

In their hour of greatest need, American healthcare betrayed Americans in pain, feeding them en masse to money-hungry opioid manufacturers. Patients in chronic pain were indiscriminately prescribed dangerous and ineffective medications, then labeled as junkies if they became addicted or overdosed.

What brought attention to the opioid epidemic was not doctors (who were complicit), academia (which was subservient), regulators (who were enablers), or politicians (who were bought). It was the body bags.

I was driving to Cape Cod with a friend for a much-needed vacation in the spring of 2021. We had been residents together and reminisced about our training as we drove by idyllic beaches and quaint New England towns. As often happens when two doctors talk, our conversation drifted toward work.

He had just begun his job as a primary care physician when he took over the care of a patient with chronic back pain. The patient was on

OxyContin and didn't exhibit any of the stereotypical behaviors that physicians are taught characterize addiction. He never ran out of his medications earlier than he was supposed to. He never overwhelmed the clinic with phone calls demanding more pills.

Yet every time he visited my friend in clinic, he was still in pain. Gradually my friend increased his dose, and yet every few months, when the patient returned, he would still have uncontrolled pain. At the patient's request, and given his own concerns about breakthrough pain, my friend continued to increase his dose.

When the patient moved to California, the doctors there were alarmed by the high opioid doses the patient was receiving and did not feel comfortable prescribing OxyContin at that level. Yet, as they tried to cut back on the dose, the patient started using heroin instead. Not too long after, the man died of an overdose.

As he drove along, the guilt my friend felt was palpable. If he hadn't increased the dose, the patient would have been in pain and continued to experience withdrawal. Yet, as he increased the dose, the patient's withdrawal and pain only worsened.

This single story encapsulates how the opioid epidemic has evolved. As millions got hooked on prescription opioids and addiction spread through society like a virus, American streets were flooded with synthetic opioids like heroin and fentanyl, which are much more potent than their prescription counterparts. They are also much cheaper and increasingly convenient to acquire.

While deaths from opioids had begun to plateau from 2017 to 2019, averaging about forty-seven thousand per year during that period, these numbers jumped considerably during the COVID-19 pandemic. A marked reduction in opioid prescriptions, in part due to growing awareness among doctors about the dangers of the pain-killers, meant that many users turned to illicit drugs. The economic downturn, massive unemployment, forced homelessness, and social isolation brought about by the pandemic all helped reignite the opioid epidemic.[30]

Yet the greatest failing of opioids, beyond morbidity and mortality, is their inability to do anything about the momentous problem they were supposed to cure: the scourge of chronic pain. Even as the United States consumes more opioids than any other country on earth, rates of chronic pain appear to be no different than in comparable countries. What's more, those rates have not changed as opioid use has increased.

Opioids have also ignited a war between patients with chronic pain and their doctors. Despite the data showing opioids to be ineffective for people with chronic pain, many patients swear by opioids as the only intervention that offers them relief. And there certainly are patients with chronic pain who receive relief from these drugs. Many patients using opioids feel that their illness has been criminalized, that their doctor visits feel like an interrogation, where every move is scrutinized and the only things missing are a lie detector apparatus and a huddle of men in suits peering from behind one-way glass.

The overuse of opioids has put doctors in a bind. As we increasingly recognize that opioids are neither effective nor safe for most patients with chronic pain, our profession has already prescribed these medications for millions who are now trapped in a vortex of pain followed by momentary relief, leading to intense withdrawal and escalating tolerance and dependence. For patients with chronic pain on opioids, whose internal endogenous opioid system has been left in shambles, there is often no end in sight.

Pain has always been subjective. It has always been everyone's own. While scientists are working day and night to develop tests to create an "objective" means to assess the intensity of pain, such an endeavor is bound to fail. We will have an objective test for pain the same day we develop one for misery, for joy, for anger, or for euphoria. Every moment of pain we feel is shaped by our past and will shape the pain we feel in the future. Every emotional trauma that we endure as children—be it losing a loved one, enduring the fracture of a family, witnessing violence or economic hardship, having a parent incarcerated, or living with someone who is struggling with a mental health or substance-use disorder—increases the chance that we will experience chronic pain as adults. In fact, any one of these adverse childhood experiences increases the chance of developing persistent pain later in life by 60 percent. For people who endure four or more of them, the odds of developing chronic pain nearly triple.[31]

As each of us writes our story, our body pens its own. Its past shapes the present; the present defines its future. Too often the only language our skin and bones know is pain.

Every human body maintains its internal harmony with meticulous care, whether between acids and alkalis or between stress and salience. Opioids don't just tip the homeostatic scale. They break it into tiny pieces. They take a person's ability to manage their own pain, their own worries, their own fears, and their own joys and desires away from them. Some individuals can put the scale back together—if they can get off the opioids and endure the wrenching pangs of withdrawal and the intractable pull of addiction. For others, the scale can never be repaired.

The personal nature of pain can be an invitation to inhabit someone else's inner world, to witness that universe with another pair of eyes and ears, another heart and mind. In a profession overrun by technology, the assessment of pain requires doctors to use their most primitive tools, their most human senses.

Yet even a cursory reading of history or stroll through a hospital ward will reveal that the management of pain lays bare our most pestilent prejudices. Infliction of harm is the most ancient instrument of power. And nowhere are the power imbalances between people more apparent than when it comes time to evaluate their pain.

8.

THE PAIN OF THE POWERLESS

Injustice and the Arc of Suffering

The things that tormented me most
were the very things that connected me with all the people
who were alive, who had ever
been alive.

—James Baldwin

PHYSICIANS ARE IMPERFECT CREATURES. They are prone to impatience, aren't trained to work in teams, and hate being told what to do. They are, in other words, human. Yet, early in my career, I strongly believed that when a patient is in need, doctors will not discriminate against them based on who they are, what they do, or what they look like.

I was wrong. Doctors and nurses carry the same biases that their communities and their patients do. And at no point are these prejudices more visible than when we are treating a person in pain.

A very simple research study confirmed this. In the long, gnarly labyrinth that is our intestinal tract, there is a small outgrowth, like an alley jutting off the winding freeway. This dead end is the appendix, an organ whose purpose nobody fully understands. While it is now hypothesized that the appendix may serve as a reservoir for "good bacteria" in our bodies, most often the appendix only gets any attention when it is about to burst open.

When the appendix gets inflamed, it becomes swollen; if left untreated, it can explode, contaminating the sterile abdomen with fecal matter. This can be a fatal event, but it is thankfully rare. Why? Because an inflamed appendix produces the most severe pain many people experience in their entire lives. This acute pain has a purpose, for it focuses our complete attention on the problem before it's too late. This is precisely why acute appendicitis is one of the most common causes of abdominal pain among patients coming to the emergency room. In the majority of cases, treatment is actually quite simple: you cut the angry appendix out. Before patients go to the operating room, though, it is recommended that their pain be controlled with medications to keep them comfortable.

I have argued that pain is a complicated experience and that it is often unclear how well painkillers can address it, particularly when there is uncertainty about a diagnosis or how long a patient will need to take the medication prescribed. So many illnesses are characterized by ambiguous subjectivity. But acute appendicitis is quite simple: it is a common, excruciating, and easily and "objectively" diagnosable condition.

And yet, in a national study of a million children who came to the emergency room with acute appendicitis in the United States, Black children were five times less likely to receive opioid painkillers than white children. This wasn't because Black children were in less pain, were older or younger, were male or female, were deemed to be less acutely ill, or had a different type of insurance. The only reason Black children with acute appendicitis were so much less likely to receive painkillers, it seems, was that they were Black.[1]

Interestingly, nonopioid painkillers such as acetaminophen or ibuprofen were prescribed at a similar rate to white and to Black children. The authors of the study posit that the decision to prescribe an over-the-counter painkiller versus an opioid is ultimately about trust: "Prescribing an opioid requires more trust of the patient by the physician." In contemporary clinical medicine, particularly when the pain is acute, an opioid prescription is often a physician's expression of sympathy, a ratification of another's infirmity. Why would physicians not trust a Black child's anguish?

Pain and its treatment have always been about the assertion of power. Power has long been wielded through the infliction of pain, but even more so in its acknowledgment. The pain of the privileged has mattered more than the pain of the dispossessed. Whenever there is an imbalance of power between two individuals, such as that between a rogue police officer and a Black man gasping for breath or between a physician and a person writhing in pain, there will be an imbalance in how suffering is recognized and, ultimately, relieved.

Furthermore, the sickening biases that underlie imperialism, racism, and sexism have long infiltrated our biological understanding of pain. Hateful ideologies found respectability cloaked in the language of dispassionate biomechanics. Foundational pain science provided prejudiced ideas with a veneer of validity, shaping cultural notions of how humans hurt. And the compartmentalizing of pain into a mechanical process, divorced from its cultural and historical influences, made it even more likely to be misinterpreted when it afflicted the most vulnerable.

The ugly and inconvenient truth is that just as racism and sexism continue to thrive in our streets and workplaces, they also persist in our emergency rooms, clinics, research labs, and hospital wards. And perhaps no medical condition has a darker political history than pain.

The influence of pain on human behavior is so deeply embedded that it likely predates human society. "The savage discovered for himself the punishing element in pain," writes George Scott in *The History of Corporal Punishment* (2013). "It was perhaps one of the first things that savage man did discover. Pain, he found out, whenever he himself experienced it, seared his brain with a lasting impression, an impression that persisted far longer and far more distinctly than did any other impression."

Human civilization institutionalized the application of agony, using it to corral the public within its predetermined margins. Punishment for behaviors deemed criminal or deviant from social norms has often been dealt through the deliverance of physical pain. Flagellation, with either metal rods or whips, has been a central instrument of religious law in Judaism, Christianity, and Islam, among other faiths. Transgressions as varied as cursing, having sex with one's wife in public, falling asleep while on guard duty, or being unable to pay back a debt could all result in public whipping. Pain was used as retribution, but it was also intended to compel all those who witnessed such punishments to stay in line.[2]

Religions have frequently used the fear of pain both to establish hegemony over their own followers and to stifle competition. The most enduring example of how religions use painful imagery to shape human behavior is the image of Jesus nailed to the cross. Physical reprimand is a central tenet of parenting in the Old Testament book of Proverbs, which says, "He that spareth his rod hateth his son: but he that loveth him chasteneth him betimes" (13:24).

Even as the hold of religion eased in Western countries after the Renaissance, many secular institutions that replaced the theological establishment eagerly incorporated corporal punishment, clothing it in the veneer of law. In *An Essay Concerning Human Understanding*, philosopher John Locke wrote, "Pain has the same efficacy and use to set us on work that pleasure has." Courts across Europe, from England to Russia, allowed the torture of individuals for all manner of crimes; disciplinary flogging was common in the British and American militaries and the US Navy through at least the nineteenth century. While it has been outlawed in most of the world, in a few countries lashing remains a common sentence.[3]

The administration of pain as an expression of power was so common in history that until recently it was a cornerstone of how courts separated facts from falsehoods.

In medieval Europe, far from being considered tainted, confessions extracted via torture were considered the "queen of proofs." Ancient

Greeks, for their part, enacted the practice of *basanos*. *Basanos* was a dark stone used to test the purity of gold, which would leave a particular mark on the stone when rubbed against it. *Basanos* also became the name for the practice used to test the purity of one's words: physical torture. Greek thinkers, such as the statesman Demosthenes, thought that confessions obtained under the pressure of pain represented the cleanest of truths. Demosthenes claimed that "torture [was] the most certain of all methods of truth," especially when applied to slaves, since "of slaves put to the torture no one has ever been convicted of giving false testimony."[4]

Torture also served another purpose for the Greeks: it established a hierarchy among men. "Torture underscored and cemented the strict separation of society into free men and slaves," writes medical historian Giovanni Maio. "Torture was thus an 'instrument of demarcation,' that drew a distinction between the untouchable body of the free citizen and the worthless body of the slave." Torture was particularly effective at eliciting confessions from slaves because, as Aristotle argued, unlike their masters, who could reason their way to deciding whether to reveal or conceal the truth, slaves lacked the ability to reason and thus would always be coercible when mutilated.[5]

The use of torture, often prescribed by courts to extract confessions in the absence of other evidence, became so pervasive in medieval Europe that it even spread to cultures in which torture had traditionally been taboo. What allowed torture to become so pervasive and acceptable was that neither the infliction nor the experience of pain was seen as necessarily reprehensible—far from it. Pain was a cosmic power and the key to God's presence within us. Pain was the road to redemption, as much for the saint as for the sinner. And therefore medieval Europeans saw torture as a way to both extract God's truth from within the human body and to initiate the criminal's salvation.[6]

Through much of human history, pain was not a repulsive sensation to be avoided at all costs but a transcendent experience at times akin to a spiritually cleansing bath. We didn't just flagellate others; we often flagellated ourselves, frequently as part of religious rituals. In some religions, such as the Shia sect of Islam, self-mutilation with whips and blades connects the devout to the suffering of long-lost ancestors.

For these reasons, historian Lisa Silverman has posited that torture was not only accepted in medieval Europe but encouraged. And it was precisely as our view of pain became more medicalized—when it went from having a point to being essentially pointless—that society's acceptance of physical torture also began to wane. And as the very thought of pain has become reprehensible, torturers have shifted toward sensory deprivation, chemical manipulation, and other techniques.[7]

Yet, even as medicine's allopathic take on pain might have contributed to a decline in its legal application as a means for institutions to exert their authority, doctors have, throughout history, facilitated this practice, and some still do. Doctors often helped monitor and optimize the administration of torture, ensuring that victims didn't die so as to allow its continuation. If victims developed injuries, physicians often treated their wounds only to allow their punishment to resume.

Not only did doctors often abet brutality, as documented in historian Joanne Bourke's 2014 book *The Story of Pain*, they also played a leading role in crafting bigoted and damaging myths about the way powerless people experienced pain.

Imperial London was home to people from all over the world, many of whom suffered silently or loudly in hospitals and infirmaries. To the eyes of an English medical student in 1896, "Jews, Turks and Heretics mingl[ed] together in one seething mass of injured and diseased humanity." This medical student's writing reflected the accepted and legitimized views of his time. No matter how the weak responded to pain, they remained inferior. When a Black man was stoic in the face of an injury, a surgeon in North Carolina wrote in a medical journal in 1914 that it was because "the Negro submits to pain with resignation, his sensibilities being less acute than those of a more highly-wrought nervous nature."[8]

At the same time, an exquisite sensitivity to pain was the hallmark of a supposedly civilized person because, as founder of the eugenics movement Francis Galton noted, there was a "rough proportion between sensibility and intellectual development," which explained how "savages will undergo [with] equanimity tortures which no civilized man could endure." Silas Weir Mitchell, the pioneer of phantom limb pain, captured the mood of that time, writing in the *Journal of the American Medical Association* in 1892 that through "the process of being civilized we have won...intensified capacity to suffer...[T]he savage does not feel pain as we do: nor as we examine the descending scale of life do animals seem to have the acuteness of pain-sense at which we have arrived."[9]

On the other hand, the expression of pain, when the one expressing it was deemed unworthy, was a sign of moral failure and weakness. Jews and Irishmen "made the most noise on the operating table," wrote a surgeon in the *British Medical Journal* in 1929, adding, "The Hebrew cried out through fear that if he failed to attract full attention he might miss some of the benefits of hospital care; while the Irishmen called loudly upon God and the saints, and wept and groaned because he was an emotional being to whose nature the repression of feeling, whether pleasant or painful, was foreign."[10]

European physicians made no secret of their prejudices. But it was in America that doctors weaponized their positions of power over people in pain, providing pseudoscientific legitimacy to the burgeoning colony's most heinous practices. This dark history continues to influence the treatment of hurting people to this day.

The most powerful nation on earth, the United States of America, was built on the back of one of human history's most reprehensible institutions: slavery. As enlightenment slowly spread throughout Europe and the American North in the nineteenth century, slavery remained deeply embedded in the southern states, which had built up a thriving agricultural and industrial economy supported by forced labor. As criticism of slavery swelled, many in the South turned in every direction to find justifications for their cruelty. When it came to finding physicians to abet them, they didn't have to look too hard.

Samuel Cartwright (1793–1863) trained in medicine under the mentorship of Founding Father and "American Hippocrates" Benjamin Rush, attended the University of Pennsylvania School of Medicine, and as the "Professor of Diseases of the Negro" at the University of Louisiana, New Orleans, became the medical and scientific face of white supremacy.[11] In the years leading up to the Civil War, Cartwright was able to wield a tremendous amount of influence, finding enthusiastic audiences on both sides of the Atlantic for absurd and virulently racist theories about Black Americans. And he had his mentor, Benjamin Rush, to thank.

Rush was the most influential American physician of his time and is considered the father of American psychiatry. A signer of the Declaration of Independence, he was also one of the most vocal opponents of slavery among the white elite. Yet he still pathologized Black Americans. Rush believed Black skin was the result of a leprosy-like disease called "Negroidism," which he felt constituted a disability. In a speech to the American Psychiatric Association, he implored, "White people should not tyrannize over blacks, for their disease should entitle them to a double portion of humanity." Yet such ideas emboldened people like Cartwright, who was convinced that the same pigment that colored Black Americans' skin imbued "every membrane and muscle, tinging all the humors, and even the brain itself, with a shade of darkness."[12]

Some of the conditions Rush defined were easily co-opted by overt racists like Cartwright. One such condition was "anarchia," a "species of insanity" that only afflicted US citizens, caused by "the excess of the passion for liberty." Cartwright morphed this into "drapetomania," a condition "that induces the negro to run away from service." Drapetomania occurs "if the white man attempts to oppose the Deity's will, by trying to make the negro anything else than 'the submissive knee-bender' (which the Almighty declared he should be), by trying to raise him to a level with himself, or by putting himself on an equality with the negro." To prevent drapetomania, Cartwright prescribed "whipping the devil out of them."[13]

To further dehumanize Black people, Cartwright invented "a disease not heretofore classed among the long list of maladies that man is subject to," which he called "dysaesthesia aethiopica." This concocted diagnosis resulted in a "partial insensibility of the skin," resulting in a Black person resembling "an automaton or sense-less machine." Cartwright claimed that this was a very common condition, particularly among Black people who were free and there-fore did not have "some white person to direct and to take care of them." To Cartwright's dismay, this condition caused enslaved people to be "insensible to pain when subjected to punish-ment."[14]

Cartwright propagated the idea that enslaved Black people were oblivious to pain and that the treatment for this insensitivity was more lashing. In fact, the wounds that many slaves developed due to whip-pings were a physical symptom of the condition, claimed Cartwright, who often leaned on the Bible to justify the idea that Black people were inherently inferior.[15]

Though the medical community would prefer to ignore it, the fact is that Cartwright's ideas have unmistakably shaped American medicine and continue to do so. A study from 2016 performed at the University of Virginia surveyed the general population, medical students, and medical residents to see how many believed some myths regarding differences between Black and white people: 58 percent of the general population and more than 40 percent of first- and second-year medical students believed that Black people had thicker skin. While the proportion was lower among senior medical students and residents, almost one in four retained this belief. Medical students and residents who ascribed to such false beliefs were also less likely to accurately perceive the intensity of Black patients' pain and recommend appropriate treatment. The persistence of such false beliefs about supposed biological differences in Black people suggests that an uninterrupted line connects the legacy of people like Samuel Cartwright to the Black person walking through the doors of a hospital in modern-day America.[16]

Funmi Akinlade was in her fourth year of medical school when she rotated to an emergency room in the eight-thousand-person town of Grand Blanc, Michigan. Besides her, there was only one other physician in the emergency department who was Black. Akinlade went to medical school in Nigeria; this rotation was a rare chance for her to work in an American hospital.

Approaching the end of a shift, she picked up the medical chart of a patient who had fallen down the stairs at his home two weeks earlier. He had been to other emergency rooms since then. An X-ray of his chest taken on the day he fell didn't show any abnormality. Yet his pain had never relented.

Akinlade walked into a scene that horrifies her to this day.

"He was really struggling and could barely speak," she told me. "His kids were scared; his wife was terrified."

As soon as he saw her though, he exclaimed, "A Black doctor—thank God."

"I had never met anyone who was so grateful that I had walked in there," she told me.

Akinlade immediately thought that the patient might have fractured his ribs and wasn't reassured that the initial X-ray didn't show anything unusual. X-rays are notorious for missing rib fractures, especially when performed soon after the injury. She wanted to get a CAT scan. "I couldn't understand how that was something potentially overlooked," she said.

In that moment, she felt her role shifting from physician to advocate. Before she went back to her supervising doctor, she went on UpToDate, a medical education website, just to make sure she would have the proper reference to back up her recommendation, but to her surprise the supervising physician acquiesced anyway.

"A part of me was hoping I was wrong," she told me. But she was right. The patient had three fractured ribs. If it were not for her, nothing might have been done for this patient's pain.

Pain and pain-related disability are both more common among people who identify as Black. Yet Black patients are routinely treated less intensively by physicians for their pain. Black patients are less likely to receive surgery for a displaced disc in the spine or to receive painkillers after undergoing surgery. And nowhere is this difference more apparent than in how doctors choose to prescribe them opioids.[17]

One irony of conventional American medicine is that despite the dangers posed by opioids, their prescription often represents a pact of trust between a doctor and a patient. The prescription expresses a doctor's belief both that the patient's pain is real and that the patient will not become addicted to the drug. It is a medicalized endorsement of a patient's integrity and character, backed by the doctor's signature and Drug Enforcement Administration number. When it comes to opioids, many doctors just don't trust their Black patients enough.

When patients experience agony from conditions that show up on tests or X-rays, such as fractures, there is no difference in how likely doctors are to prescribe opioids based on a person's race. Yet when it comes to discomfort from conditions that might not be diagnosable based on a supposedly objective test, such as a migraine or back and abdominal pain, Black people are much less likely to be given opioids. A side effect of this trend is that between 1993 and 2010, as deaths from prescription opioid use skyrocketed among white people, deaths from opioids remained largely unchanged among Black individuals.[18]

The underprescription of opioids among Black people was not just a sign that doctors were less likely to trust Black patients; it was a calculated product of structural racism. Even though Black Americans have similar rates of illicit drug use as white Americans, they are incarcerated six to ten times more frequently for drug-related offenses. The response to the prescription opioid epidemic, which overwhelmingly affects white communities, has been christened a "White drug war" that "has carved out a less punitive, clinical realm for Whites where their drug use is decriminalized, treated primarily as a biomedical disease, and where White social privilege is preserved."[19]

Indeed, OxyContin was deliberately positioned as a "white opioid" and heavily marketed in largely white, often rural or suburban communities. One white teenager in Philadelphia who was addicted to OxyContin told a reporter, "It's weird, because the kids in my neighborhood think if you are on heroin, you're a junkie. You're no good. You're the filth of the earth. If you do Oxys, it's not that bad."[20] Heroin, of course, is often associated with urban residents of color.

This strategy proved to be a commercial success but a public health catastrophe. "While many opiates have been introduced to the market as non-addictive pain relievers or even cures for opiate addiction, including heroin in 1898, no opiate has managed to stay a medication and concomitantly no opiate has managed to stay White," write Julie Netherland and Helena Hansen. "The very racial segmentation of markets into licit and illicit, White and Black, clinical and recreational as dictated by the War on Drugs and by the profit imperative of opioid manufacturers helps to drive cycles of demand and sustains a moving target of time-bound patents on new technologies of bioactive molecules and delivery devices."[21]

Still, despite the prevalence of OxyContin on the black market, its status as a "white" drug in contrast to heroin continues to elicit a gentler response from both law enforcement and the public. Even when prescription opioid use was surging, rates of arrests for heroin possession were five times greater than for possession of synthetic opioids. Interventions that treat addiction like a disease rather than a sin overwhelmingly target white people. The drug Suboxone is one of the few therapies shown to reduce risk of death among opioid users. Yet the vast majority of people prescribed Suboxone are affluent whites, and it is rarely even available in communities that are predominantly Black or Brown.[22]

Still, the fact that Black people are not prescribed opioids as frequently as white people does not explain why Black people feel pain more frequently than white people and are more likely to be disabled by it. A vast amount of research has shown that Black people are more sensitive to pain than white people, even when it is being delivered with a pin or a hot instrument in an experimental setting.[23]

Many studies have been devoted to figuring out why Black people have a heightened response to painful stimuli, both in the laboratory and in the real world. Much of the focus has been on identifying biological differences that might explain this variance. Those leading the search for this "hot spot" of pain in Black people's brains include Tor Wager, a professor of neuroscience now at Dartmouth College, who in 2013 identified the neurologic pain signature, a pattern of brain activity that lights up on functional MRI (fMRI) scans in response to heat pain. This signature tracks closely to nociception, the purely physical response to a hurtful stimulus.[24]

To study differences in how white, Hispanic, and Black Americans experienced pain, Wager recruited individuals from all three groups and applied a heating instrument to their forearms while their brains were being scanned using fMRI. The study, published in *Nature Human Behavior*, found that Black people rated the same temperature as much more painful than white or Hispanic people, whose ratings were essentially the same. But when Wager and his graduate student and the study's primary author, Elizabeth Losin, looked at the neurologic pain signature, it was similarly intense across all three groups. This suggests that there is nothing intrinsically different in how nociceptive signals are detected or transmitted among people of different races and ethnicities.[25]

Other parts of the brain, however, outside the regions associated with the physical experience of pain, were more active in Black subjects. These parts of the brain, the ventromedial prefrontal cortex and the nucleus accumbens, are known to be associated with emotional regulation. While not involved in acute pain, they are more active in people experiencing chronic stress and chronic pain. These parts of the brain did not light up as brightly when white or Hispanic participants were prodded.[26]

To account for this difference, the researchers measured nineteen sociocultural factors known to be associated with how intensely people might experience pain. These included low income, stressful life events, discrimination, and hypervigilance, among others. Of all the factors, they found only one that mediated the relationship between pain intensity and ethnicity: a history of the participant responding to racial discrimination. In fact, a history of responding to discrimination was highly correlated with increased activation in parts of the brain associated with emotional regulation and chronification of pain.

It makes sense that a history of perceiving discrimination would make a person more vigilant and suspicious of those around them, particularly when at least one of those people is wielding a hot instrument. Being discriminated against might also make people more worried that their agony will not be attended to, that they will be disbelieved and distrusted, and that they will be abandoned to navigate their suffering by themselves. Research suggesting that Black Americans are more likely to catastrophize about their pain may be partly explained by these factors.[27]

There was one last variable though. A white male researcher in his thirties had conducted all the evaluations in the study. So in a subsequent experiment, Elizabeth Losin looked at whether the results would change if the researcher were of a different race. This follow-up experiment showed that Black people had reduced pain when the clinician was also Black. No such effect was noted for white participants.[28]

Studies that use fMRI are notoriously difficult to replicate, and many studies investigating racial differences in pain sensitivity include few patients, making it likely that the findings are a product of random chance. Yet the link between a perception of being racially discriminated against and experiencing more pain is well established in studies performed both in the lab and in clinical care.[29]

This makes the role of people like Akinlade and other Black doctors even more important. Akinlade sees herself as "a bridge between the patient and equitable health delivery," she told me. "I might be the difference between the patient getting great healthcare and the patient getting suboptimal care."

Only about 5 percent of practicing American doctors are Black, a proportion that hasn't budged much over the past few decades. That relative scarcity fundamentally alters Black doctors' experience of practicing medicine. "People want me to be exceptional, when I just want to have a good day," Akinlade said. "I don't want to have the weight of the entire community on my shoulders."[30]

Akinlade feels this especially strongly as she is both Black and a woman. One of the other hues that colors the story of pain is sexism. Pain and painful conditions are far more common among women. Yet the mistreatment of pain has been one of the most enduring applications of sexism in the annals of human existence.

When Terry Gross interviewed Toni Morrison in 2015, the conversation turned to Morrison's increasing disability. She was eighty-four at the time and had undergone back surgery, but it only helped her for about eight months, and now she was switching between a wheelchair and a walker.

"Do you feel like your body hates you?" Gross asked.

"I did so much for you, body. Why aren't you helping me now? I was so nice to you," Morrison said.

"It's hard to make peace with your body when you are under attack with pain," said Gross.

"I do feel like I am under attack," Morrison replied with a laugh.

Modern medicine has helped us live longer than ever before, and women have especially benefitted from this seismic shift. While in the nineteenth century, men and women on average both had a lifespan lasting into the mid-forties, now on average women live five years longer than men.[31]

Yet women also live longer with disability and are much more likely to experience chronic pain. Women experience pain more frequently, more severely, and more persistently than men. Not only are women more likely to have chronic pain in the back, in the joints, and in the neck, but they are also more likely to have almost any condition associated with chronic discomfort, including migraines, fibromyalgia, irritable bowel syndrome, and rheumatoid arthritis. Given that many of these painful conditions that predominantly afflict women do not show up on an X-ray, as a fracture would, or in a blood test, they end up becoming what have been called "contested illnesses," with clinicians often the ones doing the contesting.[32]

There is an additional, more surprising reason why chronic pain is more common among women. Decades of research in the lab have revealed that women have a lower pain threshold and are more sensitive to pain than men. Whether pricked with a pin, prodded with an electrical probe, exposed to a caustic chemical, or asked to hold a hot object or dunk their hand in freezing water, according to the majority of studies, females experience more pain more quickly than males. An analysis of eighteen studies, in fact, showed that women are even more likely to experience pain from taking a placebo pill. This pattern is also true in other species: 85 percent of studies show that female rodents are more sensitive to pain than males.[33]

Societal gender roles, the actual biological differences between the sexes, and the meaning we give pain and how it relates to our body all contribute to how women feel pain and how they are treated when they do. And of all the pains a human being can experience, perhaps none is more mythologized, theologized, or politicized than the pain archetypally associated with women: the pain of childbirth.

Childbirth is considerably more difficult for humans than for other species, which rarely exhibit the pain behaviors humans do. Why? Our large brains may make us smarter than most animals, but they make for a treacherous passage through the female pelvis. The carbohydrate-rich diet we have enjoyed since the advent of farming may have also made our fetuses larger. For these and other reasons, human babies take much longer to emerge from the uterus: human labor on average lasts around nine hours, compared to just two for apes, and weak evidence suggests maternal mortality is higher among Homo sapiens compared to prior hominids.[34]

So not only does childbirth hurt women, but that pain has been aggravated by the cultural meaning ascribed to it in European society, set by the first woman to give birth ever. Allegedly.

When Eve succumbed to the charms of the snake and ate the forbidden fruit, God was furious: "To the woman he said, I will make your pains in childbearing very severe; with painful labor you will give birth to children." For centuries the pain of labor was understood as divine retribution for women's intransigence. Women themselves were the reason childbirth was so excruciating.

This belief is not a Christian innovation, although the Christian version is certainly severer than that of other major religions. In Islamic scripture, for example, Adam and Eve consume the apple jointly. Together they seek compassion, and God forgives them. This might be why, even though the endurance of labor is deemed to have spiritual value, pain relief during childbirth has never been criminalized in Islamic culture.[35]

While giving birth was already perilous enough, with more than one in a hundred deliveries resulting in the mother's demise, the overwhelming influence of Christianity in Western Europe meant that agony was also a hallmark of almost all births. Painful delivery was the natural order of things, while painless delivery was the work of the devil, and attempts to make childbirth bearable were often punished with death.[36]

Such was the fate of Agnes Sampson. In 1591 in Edinburgh, Scotland, Euphemia Maclean, the daughter of a lord who had married into wealth, turned to Sampson, a midwife and medieval healer, to ease the passage of her twins. As instructed, Maclean laid a bored stone under her pillow along with bits of fingers and toes from corpses. One of her husband's shirts was placed under the delivery bed. Although the twins were born healthy, they would know little of their mother. Sampson was the first woman to be burnt for practicing witchcraft in Scotland, followed soon after by Maclean in Edinburgh, at the direct command of King James VI.[37]

For most of time, women's pain was within the purview of domineering men, even as the actual work of delivering babies was handled by women. Obstetrics textbooks were written by men who had never attended a delivery. When one aspiring German gynecologist in Hamburg, Dr. Wertt, attended a delivery dressed as a woman in 1522, he was burnt alive for it.[38]

So childbirth went for hundreds of years, until one of the most monumental advances for women's rights came to pass, just fifteen miles from where I live.

On October 16, 1846, in an amphitheater now called the Ether Dome, a young painter held a mouthpiece, breathing in ether gas until he fell asleep. William Thomas Morton, a local dentist who had orchestrated the event at the Massachusetts General Hospital's operating room, made an incision in the patient's neck to begin removing the tumor that had grown there. The painter didn't even wince and, when he woke up afterward, proclaimed that he had felt no pain throughout the operation. Thus took place the first demonstration of general anesthesia.

In one simple operation, surgery was transformed from a brutal and harrowing scene to one of precision and sterility. It was a quantum leap in humanity's burgeoning quest to stave off not just death but now discomfort.

As welcome as anesthesia was, it was a rebuke to the supposed sacredness of pain. If God gave us pain, there must have been a reason for it. What might be the consequence of taking pain away, making it optional rather than predestined? Even so, physicians around the world adopted anesthesia with uncharacteristic enthusiasm. And when word reached one of the most forward-thinking physicians in Europe, he took aim at the most sacred pain of all.

James Young Simpson's mother bore seven children before she had him. She died when he was only eight. When Simpson's preternatural intelligence became apparent, his family pooled money to send him to the University of Edinburgh, where he initially studied the arts and humanities but later switched to medicine. In 1840, at twenty-eight years of age, he became a professor in a field he might have been killed for practicing in not too long before: midwifery. The chair in obstetrics and gynecology to which he was appointed had been established in 1726 and is considered by many the oldest in history.[39]

Simpson pushed for birth to be painless, vigorously attacking the high status that society had afforded to the supposed utility of pain. He quoted the Greek physician Galen: "Pain is useless to the pained." His advocacy broadened the very role and responsibility of the physician. "The saving of human suffering," he wrote, "implies the saving of human life."[40]

In 1847, only four months after Morton used ether in Boston, Simpson used it to help a woman with a severely deformed pelvis deliver a breech baby successfully. This was the first baby ever delivered with anesthesia. Simpson was not satisfied with ether, however, which irritated the lungs and was slow to take effect. In his search for a better anesthetic, Simpson often spent his evenings at his home with friends, trying different chemicals, hoping one would put them to sleep, until one day he woke up in the morning surrounded by other guests snoozing from the previous night. The substance they had inhaled the night before was chloroform.

After orchestrating the first delivery with ether just a few months earlier, in November 1847 Simpson assisted a woman whose previous labor had lasted three days, resulting in the head of the infant tearing open. For her second delivery, she became the first person ever to receive chloroform. "Shortly afterwards, her infant was brought in by the nurse from the adjoining room," wrote Simpson in a case report, "and it was a matter of no small difficulty to convince the astonished mother that the labour was entirely over, and that the child presented to her was really her 'own living baby.'"[41]

The use of chloroform spread like the plague throughout Europe. Charles Darwin gave it to his wife, Emma, after she demanded during labor, "Get me the chloroform!" Yet its most influential user was Queen Victoria, who in 1853, on the occasion of the birth of her eighth child, Leopold, inhaled chloroform for just under an hour, calling it "blessed chloroform, soothing, quieting and delightful beyond measure."[42]

Some doctors fought this revolution. One American physician preached about how wrong it was to interrupt the "natural and physiological forces that the Divinity had ordained us to enjoy or to suffer." Yet Simpson loudly chided holdouts. "Those who most bitterly oppose [childbirth anesthesia] now will be yet, in ten or twenty years hence, amazed at their own professional cruelty. They allow their medical prejudices to smother and overrule the common dictates of their profession and humanity."[43]

The development of anesthetics and painkillers such as morphine toward the end of the nineteenth century coincided with what is now referred to as the first wave of feminism. In America, where most women continued to experience "the antique burden of a suffering which the other half of humanity has never understood," first-wave feminists demanded a treatment used to alleviate labor pains in Germany. Yet their forceful demands for doctors to prescribe "Twilight Sleep" hurt the movement, not only because of the harm the anesthetic caused but because for many in this burgeoning movement, the only women whose pain mattered were those who were as white, affluent, and modern as they were.[44]

Twilight Sleep was a combination of morphine and scopolamine, a drug that dries up the membranes but, more importantly in the context of labor, causes forgetfulness. The treatment shot to fame in the

United States after two journalists, Marguerite Tracy and Constance Leupp, traveled to Germany and profiled its use for *McClure's Magazine* in 1914. In both countries, however, the movement to promote Twilight Sleep was rife with the classism that is so endemic to the assessment of pain. According to Bernard Krönig, the German physician who codeveloped the protocol, Twilight Sleep was only for the "modern woman" because to "inflict upon her many burdens and sufferings which a cruder type of woman took as a matter of course is unnatural." In that landmark *McClure's Magazine* article, titled "Painless Birth," Tracy and Leupp announced, "Modern science has abolished that primal sentence of the Scriptures on womankind: 'In sorrow thou shalt bring forth children.'"[45]

The article, the most widely read in the magazine's history, created a public fervor and led to the establishment of a national Twilight Sleep Association. Yet American physicians resisted the treatment. One physician railed against the "fallacious arguments and pictorial intimations that are being pressed on the American public in support of this very doubtful procedure in obstetrics." Leaders of the movement countered, arguing that physicians' resistance stemmed from the intensive monitoring required for the women receiving Twilight Sleep.[46]

Unable to appeal to the patriarchy's compassion, advocates for Twilight Sleep used a different approach to amplify their message to the American people by drawing on their white supremacist tendencies. The fear of childbirth, it had been argued, was dissuading white women from having children, causing President Theodore Roosevelt in 1905 to proclaim that "a race that practiced race suicide—would thereby conclusively show that it was unfit to exist." Many advocates for pain relief drew on arguments like Roosevelt's, claiming that pain-free birth would strengthen "the white stock."[47]

The campaign to promote Twilight Sleep was meant to usurp the male-dominated medical establishment, using racial purity as but one vehicle to advance that agenda. But it set women back considerably.

Twilight Sleep was associated with delayed labor, searing headaches, and moments of delirium, leading to women being confined to canvas cages during delivery. In fact, one of the most prominent proponents of Twilight Sleep, Francis Carmody, herself died during delivery from excessive bleeding in 1915. Male physicians used these complications to argue that they, rather than women, were best positioned to manage women's bodies. Once again, women in labor would have no respite from pain.[48]

Things would change yet again in the mid-twentieth century, however, with the invention of the epidural. Doctors found that an injection of anesthetic into the outer layer of the spinal cord numbed the pain of labor. While there was an initial learning curve (the first drug used was cocaine), the procedure became increasingly safe, and no less than the founder of modern pain medicine, John Bonica, used it on his own wife. An epidural does not cause women to lose consciousness and therefore lose control over their bodies; nor does it come with the side effects that morphine and scopolamine do. It is close to the perfect means of achieving a birth that is both painless and safe. My wife didn't hesitate for a second in asking for an epidural when in labor with our daughter. When Eva emerged into the world at 3 a.m. after a marathon labor, my wife was so comfortable, she was half asleep.[49]

Even so, myths about the necessity of a painful birth endure. Baseless claims routinely circulate that epidurals are risky, that they can hurt the baby, and that they can prolong labor. The development of the epidural coincided with the rise of the natural birth movement, which sought to strip modern technology from the experience of labor. To many in the natural birth movement, the pain of childbirth is of considerable significance. "In achieving the depersonalization of childbirth and at the same time solving the problem of pain, our society may have lost more than it has gained," writes natural birth activist Sheila Kitzinger in her 1978 book *Women as Mothers*. "We are left with the physical husk; the transcending significance has been drained away." Today, many continue to perceive even an innocuous saline IV in the arm during labor "as the devil."[50]

Yet, ironically, even as many progressive women continue to champion the natural birth movement, many of those who laid its foundations were misogynistic male obstetricians. The very person who helped conceive and popularize the concept of natural childbirth was Grantley Dick-Read, a British obstetrician whose real motivation was to alleviate the fear affluent and white women had of childbirth so that they would have more babies. "Woman fails when she ceases to desire the children for which she was primarily made. Her true emancipation lies in freedom to fulfil her biological purposes," he wrote, noting falsely that "primitive women" were less likely to feel pain during childbirth than Western women. Dick-Read knew exactly the purpose women served: "The mother is the factory, and by education and care she can be made more efficient in the art of motherhood."[51]

The image of the tranquil birth can often diverge from reality. Ellie Slee, a British freelance writer, was determined to have a birth free of medications. However, her plan crashed, and she ended up having to go to the hospital, where they found that her fetus had developed signs of distress. When she did reluctantly receive the epidural and immediately stopped writhing, her husband declared with "uneasy wonder" that "the epidural is modern medicine at its finest."[52]

Perhaps we can chart a middle ground, one that emphasizes that the choice to use or forego pain management in childbirth has no moral dimension. At the same time, we also need to open our eyes to the fact that not all birthing parents share the same experience or receive the same response. Even though Black women report experiencing more pain during childbirth, they are less likely to be provided painkillers. Compared to their lighter-skinned counterparts, darker-skinned Brazilian women are much less likely to receive a local anesthetic injection when a surgical cut is made at the vagina to ease the passage of the baby. In a global study that included almost 315,000 women, only 6 percent of women delivering vaginally were provided with any pain relief whatsoever. While women's pain has been underrecognized, the effects of those prejudices are multiplied by race and class, biases that are held not just by men but by many women too.[53]

Yet, while biological differences in how we perceive pain are likely scant if not absent between people of different ethnicities, the same is not true for men and women. Some research suggests that those differences might in fact change the way each sex perceives and experiences pain. While men and women are overwhelmingly similar, understanding how women hurt differently than men will be critical to relieving suffering not just for one half of the population but for all of us.

A wealth of research has demonstrated that women are generally more sensitive to pain than men. This holds true across cultures and in both experimental and clinical settings. Most people find this statistic surprising, and many might even find it troubling or offensive. Many women experience pain every month when they menstruate, and the majority of women who bear children manage to bear immense pain. And gender conditioning can help explain the apparent gap in pain sensitivity: in a culture that glorifies masculine toughness and machismo, men are less likely than women to admit to being in pain. In such a culture, saying that women are more sensitive to pain may feel like calling them weak.[54]

Evolution would beg to differ. Organisms that are more keenly aware of their bodies are more likely to detect threats, and those that demonstrate more pain behaviors are more likely to receive help from their communities. While no one would desire a heightened sensitivity to pain, it might in fact help us survive. This difference between the sexes might even be one reason women tend to live longer than men. And these differences can't entirely be chalked up to culture. Indeed, research is uncovering sex differences in the very physiology of nociception.

After thalidomide was introduced for morning sickness in the 1950s, causing severe abnormalities in tens of thousands of infants around the world, women were largely excluded from drug trials because of fears that women might be, or might become, pregnant during the studies. Yet such fears would not explain why even basic science research overwhelmingly uses tissue from male organisms. A 2005 review by Jeffrey Mogil, the scientist leading the study of sex- and gender-based differences in pain, found that 79 percent of pain studies involved only male animals, and only 4 percent considered the possibility of sex differences. Subsequently, in 2014 the National Institutes of Health instituted rules to require the inclusion of both male and female animals in basic science research. New research is now leading many to believe that we might not be far from having different drugs to treat pain in men and women. A case in point might be medication for one of the most common causes of chronic pain: migraines.[55]

Migraines affect more than a billion people worldwide. They are the second leading cause of years lost to disability among all medical conditions. And their burden is not shared equally between the sexes. Women are two to three times more likely to have migraines, and they are the leading cause of years lost to disability among women. Migraines are more impactful for women too, causing greater disruption in their lives.[56]

These differences may, at least in part, be chalked up to the effect of sex hormones in men and women. Hormones play a major part in migraines, which are similarly common among young boys and girls but spike in girls after they hit puberty, when their bodies are flooded with estrogen. Fluctuations in estrogen levels have long been implicated in migraine onset, and not only in women; men with migraines have greater levels of estrogen and lower levels of testosterone than men without migraines. In fact, research suggests that testosterone reduces sensitivity to all pain. And trans women receiving cross-sex hormones end up experiencing migraine rates similar to cis women, while trans men experience a lower risk of headaches than cis women. Therefore, hormones certainly play a part in why men and women feel pain differently.[57]

These differences have important implications for the treatments we develop for pain. Recent years have seen a new class of drugs developed for the management of migraines. These new drugs inhibit a molecule called calcitonin gene-related peptide (CGRP), found to be elevated in the blood of those having migraine attacks. CGRP levels have been shown to be elevated in a number of conditions, from osteoarthritis to soreness after exercise. Research published in 2019 showed that exposing the brain's membranes to CGRP causes female mice to exhibit pain behaviors. But this was not true of male mice, who appeared unaffected. Dig a little further and you'll find that more than 80 percent of patients who received the CGRP inhibitors in clinical trials were also females and that estrogen has been shown to augment the effect of CGRP.[58]

So even though CGRP inhibitors are no silver bullet for migraines, the fact that preliminary studies show that women are far more sensitive to this molecule points to the growing need to study how the male and female bodies in pain differ.[59]

Yet why men and women hurt differently is not just a matter of sex and biology. It is also a matter of gender, which encompasses the social and cultural characteristics that come to typify boys and girls and men and women.

When I first learned that my name, Haider, means "lion" in Arabic, I was impressed. Lions are valiant and formidable. Yet as I grew older, to my utter dismay, I realized that the name of almost every Muslim boy I met growing up in Pakistan also meant "lion." From Ali to Zafar, hundreds of Arabic names for boys translate to "lion." In contrast, I have yet to meet a single girl whose name means "lioness."

When we had our daughter, Eva, my wife and I decided that we would not pigeonhole her the way so many girls around us are. No pink clothes. No princess dolls. No kitchen sets. And yet, just a few years into her life, she is about as prototypical a girl as you will find in your local kindergarten.

Culture is the inner voice of our inner voice, and gender roles are the sharp edges of powerful societal forces that shape how we dress, how we eat, and how we think about who we are and our place in the world, even when we are all by ourselves. Gender also comes to define our relationship with our bodies, helping to translate what our bodies feel into what that feeling means. From the moment they are born, males and females are trained to react to discomfort in divergent ways that qualitative researchers have documented comprehensively.

Women are sensitive and delicate; they are taught to be careful and ask for help. Meanwhile, men are stoic and resilient; they are taught to fight through pain and not to ask for help with it. Men in pain are much less likely to seek care from either family or physicians. If pain threatens their sense of masculinity—for instance, hindering their ability to lift weights or play sports—men would rather risk injuring themselves further in order to preserve their sense of self.[60]

Women are treated differently in part because they are also much more likely to have pain from conditions that are without an obvious cause. This transforms many medical encounters into confrontations between the medical enterprise and the woman's person. In one survey, Canadian clinicians described their patients with fibromyalgia, which occurs between three and twenty times more commonly in women than men, as "malingerers, time consuming, and frustrating," with some doctors holding "the patients accountable for their pain."[61]

The narrative men carry about their pain is different from that of women. When men hurt, they often ascribe the pain to external forces. They treat it as an occupying force they resist. When women hurt, the pain often comes with self-loathing. Women are twice as likely to be depressed as men, and women who are depressed are twice as likely to have chronic pain as those who aren't.[62]

Modern medicine responded to the pain women were feeling by doing the only thing it knows: giving them drugs. Multiple studies show that women are more likely to be prescribed opioids, in higher doses, and for longer periods than men. This might be because women are more likely to seek medical care in general, for both physical and mental ailments, as well as for pain.[63]

Yet women in pain face a disadvantage when it comes to seeking that care: while women represent half of medical students and a third of anesthesiology residents, only 18 percent of pain physicians are women. Among all the subspecialties that one could undertake after anesthesiology, women are least likely to pick pain medicine. Only one of the thirty-five past presidents of the American Academy of Pain Medicine is a woman; as of November 2021, there was not a single woman on its executive committee.[64]

The story of one of American pain medicine's pioneering female physicians might offer a clue as to why this disparity exists.

I did my residency at the Beth Israel Deaconess Medical Center in Boston, which is affiliated with Harvard Medical School. The pain center there is named after Carol Warfield, a pain specialist who was previously chair of the department of anesthesia and an endowed professor at Harvard Medical School. Simply being a woman makes her the rarest of rare breeds in the upper echelons of pain medicine.

That there are so few women in pain medicine is quite odd. Pain medicine doesn't involve overnight calls like other anesthesia subspecialties, making it more feasible for those with family responsibilities. Even specialties that have historically been openly hostile to women, such as surgery, have far more women practitioners than pain medicine. The reasons for the lack of women in pain medicine might be rooted in the core of how this specialty has evolved. "The greatest potential deterrents to female participation may actually be the culture and historical trends of pain medicine," write anesthesiologists Tina Doshi and Mark Bicket from Johns Hopkins.[65]

In 1979, after she finished her clinical training, Warfield started one of the first pain clinics in the country and eventually rose to become the chief of anesthesia, which made her the most senior woman in the entire hospital. Although she was well liked within her department, her ascent was greeted with considerable misogyny. Even as she was running the clinic, Warfield was raising her three children by herself. Her husband had died just a few months after their third child's birth. "If I ever said I couldn't come to a committee meeting because of my children, I would be off the committee. If a man said that, they would say what a good father."

Matters reached a dramatic climax when one day she was unceremoniously fired from her position. The head of surgery, it turned out, didn't want to work with her. "They paid much lip service to wanting more women. They finally got one and then as soon as one of the males didn't like what was going on, they took the side of the man and not the woman," said Warfield. She was demoted from head of anesthesiology to an entry-level position in the pain clinic. "I started that clinic in 1979 but they took my name off its history."

After being approached by many other women physicians who begged her to take action, Carol Warfield sued the hospital and its leadership in 2007. She continued to work there for seven more years as the case went on, but her fate had been sealed, and she would never again achieve the level of authority she'd had in her leadership role. "When you sue your department, no one wants to do anything with you," she said. Despite the cost, she felt the move was worth it. "I was speaking for a lot of women who couldn't."

In 2013, Carol Warfield received $7 million in a settlement with the hospital, the largest gender-discrimination settlement at the time. Perhaps more importantly, she got to keep her endowed professorship, and the pain center that she had founded was renamed the "Arnold-Warfield Pain Center." "It's obviously great to have my name there, but it has been a long journey," she ruefully relayed.[66]

Gender diversity is critical for any medical specialty, but it is particularly critical for pain medicine. Women physicians are more likely to spend more time with patients, build stronger partnerships with them, and provide more patient-centered care. Female, but not male, physicians are more likely to prescribe antidepressants and make mental health referrals for their female patients with pain. This is important because almost a quarter of patients with chronic pain in the United States have depression, and women with chronic pain are more likely to be depressed than men. Yet, even as the majority of patients with chronic pain are women, men largely continue to be the judges, juries, and executioners of their agony.[67]

And women face a distinct risk due to the overprescription of opioids by the overwhelmingly male-dominated specialty of pain medicine. Research shows that women are often given opioids while also being prescribed benzodiazepines and other medications that increase overdose risk. These patterns have contributed to the doubling of opioid overdose deaths among women from 2009 to 2019. Women are also at higher risk of developing drug cravings and relapse, having greater breathing problems, and developing more severe psychiatric, medical, and employment complications from opioids than men.[68]

These statistics make clear that it's not enough for physicians to become more aware of the biases and prejudices that lead us to dismiss the pain of women or people of color. Indeed, if our instruments for relief are flawed, they could in fact worsen the gaps between the privileged and the vulnerable.

Despite the persistence of inequalities in medicine, there is reason to hope. Recent national US data from the Medical Expenditures Panel Survey indicate that women are more likely than men to report that their doctor always spent enough time with them and always explained things clearly. Among all racial and ethnic groups, Black Americans are actually most likely to report that their physicians always treated them with respect.

At the same time, though, when it comes to how doctors and nurses treat the person in pain, our biases are fairly clear, and they are perhaps most clear in who doctors believe "deserves" opioids. An opioid prescription is unlike any other we sign: it is based on an entirely subjective assessment of another person's account and, given how obvious the addictive potential of these drugs has become, requires us to make a tacit judgement of the patient's character.

Such a situation is bound to be a morass mired with our often subconscious preconceptions. With no imaging scan that can tell us how much pain someone is feeling or blood test informing us about the risk a narcotic might pose for the person, we are left to our most human instincts. These can often lead us to connect with others in a profound way, but they can also reflect deeply held implicit biases. The recognition and treatment of pain might be more susceptible to a clinician's prejudices than any other aspect of medical care.

Yet, as we confront our biases, what will that mean for the person in pain? Will it mean that more people of color are prescribed opioids? That will certainly be welcome for the Black child with acute appendicitis, the Black man with fractured ribs, or the Black woman recovering from a C-section. What about the Black person with chronic back or joint pain? Opioids are neither effective nor safe in people with chronic pain. Ironically, the very bias that causes physicians to be less likely to prescribe opioids to patients of color has almost certainly led to fewer people of color dying of opioid overdoses.

That may be changing though. Opioid overdoses surged during the COVID-19 pandemic, largely due to an increase in deaths among Black Americans.[69]

While reducing individual physician bias is essential, it's not enough to bring members of disempowered groups into a broken system. If the mission of medicine is the elimination of suffering, we cannot succeed without a broader economic, racial, and social charter. It is essential that we treat not just individual ills but those that plague our communities as well. Simply put, you cannot be a good doctor or nurse without being committed to social justice. This is unfortunately not reflected in the state of contemporary healthcare, where the management of chronic pain essentially boils down to either prescribing opioids or performing invasive procedures. While opioid risks have been widely documented, the fact is that most procedures performed to manage chronic pain are no better than placebo either. The people whom procedures for pain seem to benefit most often are the practitioners profiting from them and their employers. Subjecting vulnerable communities to contemporary pain medicine will only widen the pool of people languishing in the death spiral of relentless pain and ineffectual care.[70]

A fundamental rearrangement is needed. For most of time we believed that all affliction was a cosmic instrument of metaphysical justice, as if divine forces were reaching down and pinching individual nerves, to teach us a lesson, to mend our ways, soaking us in pain's emancipating glow. Within just a few decades, driven by a flourish of capitalistic industrial forces, pain was expropriated from the metaphysical world and transformed into an entirely mechanical process. Our rich cultural history and understanding of what it means to suffer were replaced by focus-group-generated marketing slogans garbed in science. Far from protecting their patients, even well-intentioned physicians became unwitting propagators of corporate interests. Yet, if it weren't for the opioid epidemic and the intense scrutiny that has followed, there is no reason to believe we would have changed our ways and sought real answers to help those mired in perpetual torment.

For decades now, the phrase "it's all in your head" has been flung at people to diminish their lived experience, to erode their personhood, and to render them invisible. Medical technology has only made this worse. Human suffering is real only if it shows up on an X-ray. Only then is the malady granted objectivity and the person it infests provided legitimacy. For the unwell person, a diagnosis is a gift, one that many with chronic pain never receive.

Modern medicine has much to be proud of. Basic and clinical science have delivered cures for many diseases. Pathologies once viewed as death sentences are now manageable chronic conditions. In the 1960s, almost half of people died after a heart attack. Only a handful of the few survivors were left whole enough to live a near-normal life. Now the vast majority of heart attack patients do well enough to resume their lives.

Yet modern medicine couldn't keep all of its promises. It didn't just promise painlessness; it promised that pain would be killed. Yet we live with more pain today than ever before. Incremental progress will never lead us forward. We need a paradigm shift.

Doctors and nurses need to reimagine what we strive for, what commitment we make to those who come to us in need. This will require us to unlearn many bad habits and rediscover some good ones forgotten along the way.

One of the most important things that medicine can do to improve our understanding of pain might in fact be to take a step back. The medicalization of chronic pain has constrained our view of what it means. The flattening of pain into a solely physical sensation narrowed our focus to only looking for physical causes and providing mechanistic solutions. And when people's suffering didn't appear to fit into this rigid and shallow definition, and didn't respond to our ineffectual treatments, we labeled them as fakers and frauds. Such accusations especially targeted women and people of color, becoming potent pseudoscientific instruments of sexism and racism.

The truth, however, is that pain really is all in our heads. And it is there, within ourselves, that we will find a way to overcome its crushing influence.

9.

ALL IN THE HEAD

The Future of the Body in Pain

Let me not beg for the stilling of my
   pain,
But for the heart to conquer it.
                    —Rabindranath Tagore

WRITING THIS BOOK WAS A PAINFUL experience, and I mean this in the most literal sense possible. Everyday aches and pains became agonizing body slams. As I began to see the summit crest from behind the clouds, finally beginning to integrate the thousands of research papers and historical texts I read with the dozens of patients and scientists and clinicians I spoke to, it became clear that spending a lot of time thinking about pain had begun to exact its own cost, its own pound of flesh.

One day I couldn't move my neck. Then it was my shoulder. Then it was my other shoulder. Parts of my body I barely knew existed came to haunt me in a way they never had before and never have since. The joints in my right hand swelled up, leaving me unable to hold a cup of coffee or slice a vegetable. It was also during this time that I developed painful shingles on the right side of my chest. I hurt so sharply in my left inner thigh that I literally whimpered as I got up one night to get a drink of water. The pain in my leg was so intense that I had to try multiple different maneuvers just to get out of bed, eventually rolling on my back and slipping out like a wet fish.

As I suffered, often in solitude, and all attempts at relief sequentially failed, I had a recurring thought that I could not snuff out: Was writing this book and incessantly reflecting on pain causing my random and unexplained but all-consuming and harrowing afflictions?

My body was challenging me, not just physically but intellectually, holding me to an even higher standard. One can weave a tale for others but rarely for the self. I began to rethink my own experiences. I questioned every sharp sting, every singeing spasm. Even when I was by myself, up at night unable to sleep, I examined every reaction I had to my ordeals: Was I overacting or overperforming? How much should I or could I limp? Was I ignoring a potentially serious disorder such as a blood clot? Had I developed a stress-induced autoimmune condition attacking my joints with cruel arbitrary delight? I found myself constantly ruminating about whether I should go to the emergency room or wait things out.

As I strained to recall memories of when my back first broke, attempting to make my battered recollection ever more vivid, to transport readers into the gelatinous space that my vertebral column was ramming into, I began to relive a nightmare I could not rouse myself from.

In those months following my injury, I tried everything I could to get better. Any time I was free between classes, I would sneak into the physical therapy suite, where the therapists, without charging me a single dime, would work on my back. I spent hours stretching my back, arching it like a camel and curling it like a cat, over and over again. I would lie on my back and thrust my pelvis up high, holding it aloft for as long as I could. Afterward I would bring my knees up to my chest, squeezing them in a tight hug. Frequently, I would be the last person to leave the center. Often the momentary respite I got during those sessions put me to sleep, the only rest I would get on many days.

I was desperate to get better, to convince myself that my injury would not force me to give up my dream of being a physician.

225

Yet every day that dream became harder to actualize. I went for an MRI that showed so many abnormalities for someone my age that the radiology team discussed it during their weekly conference of "interesting" cases. A few days later, with the films from my MRI in hand, I consulted with a famous spine surgeon. I would have had to wait months to find a slot for an appointment, but given that I was his student, he agreed to see me informally.

"I could do the surgery," he told me in the hallway between clinics, "but a spine that's been touched by a surgeon is never the same again."

I was afraid of using opioids because I was worried about becoming addicted. So a colleague recommended something that sounded radical at that time: ketamine. Ketamine is not habit-forming, and today many consider it one of our most promising weapons in the flailing war on pain.

Ketamine induces what is called dissociative anesthesia, disrupting the connection between who we are and what we feel, detaching our consciousness from our sensory experiences. Ketamine essentially defangs pain into nociception, separating it from the distress it causes.

Ketamine provided me a different way to inhabit myself. As I type these words, what I feel in the tips of my fingers is completely wedded to what I see my digits are doing. My mind occupies my body entirely, filling every pore in my skin, creating a unified sense of reality that integrates and synthesizes every neuron processing sensory and motor information. Using ketamine, what I felt did not correspond to what I saw, and what I saw did not correspond to what I felt. I heard music, but it wasn't melodious. My mind no longer occupied my body. I now occupied the entire space around me. When I was in my room, I was the room. When I was outside, I was the sky.

Yet as soon as it wore off, the pain came back. Ketamine is also known for causing frighteningly bad trips with enough frequency that there is a name for it: the K-hole. One time in the K-hole I began to forget everything—where I was, what year it was—until I eventually began to forget who *I* was. Frantically I started to tell my friend basic details about myself such as my name and who my parents were, that I was sorry, because I was sure I was about to descend into a permanent schizophrenic state. My back was already broken, and ketamine was now breaking my brain.

Ketamine is a special anesthetic because, unlike other powerful anesthetics and opioids, it does not slow down one's breathing. It can therefore be conveniently given outside an operating room or intensive care unit, and some of its most essential use has been on the battlefield. However, little if any good evidence suggests that ketamine actually helps manage chronic pain. All the same, ketamine clinics have started popping up around the United States and are attracting not just patients with chronic pain, many of whom are no longer being prescribed opioids, but also people with anxiety, depression, and posttraumatic stress disorder.

Ketamine's hallucinatory properties have long made it a rave favorite. In a study of electronic music enthusiasts, 40 percent tested positive for it. But ketamine is far from benign. It is known to have toxic effects on the heart, liver, and brain. Despite these risks, the business of ketamine is booming.

"The use of ketamine for chronic pain and other maladies has drawn comparisons to the Wild West," wrote one anesthesiologist, "as use of these infusions is random and poorly regulated." While initial evaluations for patients are often free, infusions are costly and rarely covered by insurance. Treatments can cost thousands of dollars a year.[1]

As the pendulum swings from overprescription of opioids to a full realization of their risks, leading to underprescription in many cases, desperation is spreading to find the cure-all elixir to replace narcotics. Ironically, the story of ketamine is an eerie echo of the birth of the prescription opioids it is being positioned to supersede, and it is far from the first drug to be repurposed for this fight or even the most popular. That distinction belongs to cannabis.

Cannabis, also referred to as marijuana, activates a widespread network in the body called the endocannabinoid system, comprised primarily of two types of receptors. Type 1 receptors are mostly found in the limbic system of the brain, while Type 2 receptors are much more widespread throughout the body, often in the immune cells.[2]

The high of cannabis largely comes from the stimulation of Type 1 receptors by its psychoactive ingredient: tetrahydrocannabinol, or THC. Yet cannabis also lessens inflammation, relaxes the muscles, alleviates itching, stimulates the appetite, dilates the lungs' passageways, and, for some, relieves pain. Many of these additional effects are driven by the nonpsychoactive component of cannabis: cannabidiol, or CBD.

Recent years have seen widespread efforts to decriminalize cannabis, and early data from Colorado suggested that 94 percent of medical marijuana users were consuming it to treat chronic pain. Yet, even as cannabis use skyrockets, high-quality science around whether cannabis can help with chronic pain remains absent. Cannabis is prone to the same type of tolerance that opioids are, and the long-term risks and benefits of cannabis remain unknown. All of this is also true of psychedelics like LSD, which are also now being considered as a remedy for chronic pain. Yet, if our deadly tryst with narcotics teaches us anything, it might be that our ceaseless quest to find the pill-that-will—be it to annihilate anxiety, depression, trauma, or chronic pain—might end up harming more than it helps. The opioid crisis is as much a failure of pill culture as of opioids themselves, a failure we can only compound further with more drugs.[3]

As we look to the future and take stock of all that we know about chronic pain, we find that the most promising treatment approaches don't involve any chemical inebriation or procedural manipulation. They involve one human being talking to another, helping them realize that the path to relief was within them all along.

Jenny (not her real name) had migraines for years, but lately they were starting to take over her life. She had to switch careers as she became a captive in her own body. Having worked through every imaginable medication for migraine, Jenny's parents brought their teenage daughter to the office of Elaine Goldhammer.

Goldhammer had been practicing internal medicine for many years, and much of her work dealt with treating patients' pain. It was also the most frustrating part of her work. "We all want a quick fix, for us, and for our patients," she told me, "and all we have are needles and pills."

Goldhammer took this feeling of helplessness in a direction few of her colleagues would ever imagine: she began to train in hypnotherapy. "The magic of hypnotherapy is that I can get to the root cause of what their disease is all about, where their brain came up with the idea of reproducing pain."

Hypnotherapy uncovered an entire unspoken layer of Jenny's migraines, hidden deep within her subconscious. "The narrative her brain created was that having a migraine, a physical sensation, made so much more sense than to have depression, which didn't make sense," said Goldhammer. "The brain is designed to keep us safe, and for people who have had significant trauma, their brains are just trying to make sense of things."

A few weeks later, when Goldhammer followed up, Jenny told her, "I can finally open the shades in my bedroom."

Admittedly, I approached Goldhammer with a great degree of skepticism. Illness and suffering have long attracted charlatans and quacks eager to exploit people at their most vulnerable. Yet Goldhammer, who was a full-fledged physician, made me reconsider many of my prior assumptions. Medical training prepares doctors well to prescribe pain medications, to select or adjust doses, and to identify side effects, but it doesn't equip many to have difficult conversations about people's suffering, fears, anxieties, and emotions. Too often, when confronted with a person in pain that has no "organic" causes, physicians feel helpless. And when we feel helpless, we are inclined to shift the blame from the disease to the patient. The pain must not be real. It must be in their heads.

"To sit there, and listen, to have them walk with it, and guide patients through their pain," says Goldhammer, "we are not trained to do this."

As for so many others, my only exposure to hypnosis had been as a stage trick. And even serious practitioners of hypnosis like Goldhammer warn of the lack of robust data guiding the practice. Yet that is changing due to the work of scientists like David Spiegel.

"My parents were psychiatrists," Spiegel told me as we spoke over the phone. "My father had learned to use hypnosis when he went off to World War II to treat pain and combat illnesses."

As a medical student at Harvard, Spiegel always felt that his interest in hypnosis made him a bit of an outcast. "I remember learning in medical school that opioids don't get you addicted. It's only street junkies who are using them to get a buzz who get addicted," he said. "I left Boston because there were constraints about hypnosis. I couldn't breathe."

Spiegel now is a professor at Stanford, helping transform hypnosis from a gimmick into a science revealing new depths of the human mind.

"Using the brain as a whole in a way that it can help people is not used enough," he told me. "It's a profound thing that people can do something to help themselves."

Spiegel has shown that people who are hypnotizable can effectively decide what to pay attention to and what to ignore. Sometimes hypnotizability is a trait you are born with, but sometimes it can represent an elaborate defense mechanism, a way to disassociate oneself from personal nightmares. People who have been abused as children, for example, can be highly hypnotizable.[4]

Pain presents a fundamental challenge to our ability to concentrate. From an evolutionary point of view, this makes sense. You would want pain to be the most arresting feeling that can come banging at the brain's door—it should demand immediate attention from the body that is experiencing it. If a stranger sticks a big needle in your arm, your body wants you to do whatever you can to pull it out immediately. But what if it's not a stranger? What if it is a nurse drawing blood? What if you are getting a flu shot? Do you react the same way? But perhaps more importantly, does it hurt as much?

"The strain in pain lies mainly in the brain," said Spiegel. "The brain is wired to interpret peripheral signals, and a big component is how the brain interprets pain."

What happens to the brain when it experiences recurrent pain? The moral of the story of the boy who cried wolf is that repeated and exaggerated claims may in fact have the paradoxical effect of numbing recipients to the import of the message. But the brain never read Aesop's Fables. Patients with chronic pain have consistently been shown to have a lower threshold for pain and to be more sensitive to it.[5]

"The big problem is that our brains don't habituate to chronic pain—we treat all pain like acute pain," says Spiegel. "And having the same type of novel reaction to pain is the real problem." The brain in fact becomes increasingly sensitive to pain as it experiences more of it.

The goal of hypnosis is not to fight pain but to distract the mind away from it. Is it the feeling of warmth that envelopes you in a hot tub? Is it the soft, cool breeze grazing your face and arms as you lie back on the beach? "You don't have to be the same way in your body. Go somewhere else and feel something else," Spiegel suggests to his patients. "When you teach people that they can modulate their experience of pain, that makes a real difference."

What type of difference? In a trial published in the Lancet, Spiegel randomized patients undergoing minimally invasive operations to receive either standard pain control, structured attention (in which a dedicated person attended to the patient's needs, listened intently, and provided encouragement), or structured attention with self-hypnosis.[6] Self-hypnosis was induced by patients simply listening to prior recordings of live hypnosis sessions.

The findings were stunning. While pain increased steadily in patients receiving standard pain control, the graph remained flat in the hypnosis group. This occurred despite patients in standard care getting twice the painkillers that patients in the hypnosis group did. While 15 percent of patients receiving standard care developed serious blood pressure and heart rate abnormalities because of the drugs they received, only 1 percent of patients in the hypnosis group did. And last, the procedure on average took sixty-one minutes to finish in the hypnosis group, while it took seventy-eight minutes in the standard care group. So despite receiving more painkillers, the standard care group developed more complications, experienced more pain, and underwent a longer procedure than the self-hypnosis group.

In additional trials, hypnosis has also been used during labor and with painful conditions like migraines, cancer, irritable bowel syndrome (IBS), fibromyalgia, and burns, to name a few. Hypnotic suggestions can also improve other aspects of living that chronic pain often affects, like sleep, activity levels, energy, anxiety, and depression. Finally, not only is hypnosis effective when guided by a trained professional, but self-hypnosis, as shown in the *Lancet* trial, practiced by individuals themselves, can be quite effective for many in extremis.[7]

Hypnosis is far from a cure-all for chronic pain. Yet it offers tantalizing hints about an unexplored dimension within us all whose potential remains entirely untapped. And as the opioid epidemic starts to come full circle, there might be no better time to pry open our minds. Hypnosis was the original way people like Sigmund Freud brought to the surface the human subconscious. As we yank it open further, the ability of the human body to heal itself is becoming increasingly apparent.

Pain has been so difficult to talk about thus far because pain feels so intensely personal that it seems impossible that any other person could possibly understand it. In her influential book *The Body in Pain*, Elaine Scarry writes, "Pain does not simply resist language but actively destroys it." Yet even as many believe it is our most inscrutable possession, one of its chief functions is actually communicating itself to others. Pain is designed to be expressed even without letters, without words, with nothing more than a grimace or a growl.

One way to understand pain better might be to dissect a sensation that is similar but simpler: itch.

Itch has often been defined by its reaction. The German physician Samuel Hafenreffer described it 350 years ago as "an unpleasant sensation that elicits the desire or reflex to scratch." Pain too is often inextricably paired with a reflex—withdrawal—and both have protective functions. The very purpose of itching is to initiate the scratch, an essential feature to protect human beings from their most prolific hunter, the mosquito, though it's also useful to dislodge other bugs or wipe off caustic fluid. While the pain withdrawal reflex rewards you with the absence of agony, the scratch reflex provides something altogether different: the feeling of having scratched, which has been called "one of the most exquisite pleasures."[8]

Yet itch can often transform into a ceaseless cycle of itching and scratching. Not only does the gratification from scratching create a vicious cycle, but scratching can increase the release of inflammatory chemicals in our skin, which further increases how prickly we become. And just as stress can increase how much pain people feel, it can also transform a transient itchy episode into a long-term state. And far from making a person more resilient, the longer someone itches, the easier it becomes for them to feel scratchy. In fact, the rush of endogenous opioids that our body gets from scratching an itch—a rush that we don't get if we just scratch a random part of our body—becomes even more intense in those with chronic itch.[9]

From a biological perspective, pain and itch are so closely related that for many years, itching was essentially thought of as a milder form of pain. This is because many pain receptors appeared to also be participating in the detection of prickly stimuli. Some receptors, such as TRP, which can detect heat pain, can also detect itching, though we know now that there are also some unique pathways that itching can take in our nervous system.

Despite the differences in the two sensations, itching is the closest thing we feel to pain, confirming much of what we now know about pain. Merely thinking about itching can wholly reproduce the sensation in our brain. The mere belief that a random ointment will relieve an itch will reduce how much we scratch. And far from being more discerning, those with chronic itching are even more likely to be suggestible to feeling itchy.[10]

Just as writing this book caused me to hurt, this particular section also caused me to itch, leading me to frequently rub my earlobe, scratch my elbow, scuff my chest, and flick the tip of my nose. It has been downright distressing. Yet all this is just the visible tip of the iceberg, for as pleasurable as the act of scratching is, I have restrained myself far more than I have caved. Many calls for relief from the back of my calf, the bottom of my heel, even from the nether regions, have been left to fend for themselves and resolve on their own.

While we have all been led to believe that there is an insurmountable wall between our bodies and minds, delineating what's real and what's simulated, the science of itching and pain reveals that the mind-body dichotomy is a sham, a relic of an ancient time when we understood little about how the human form was integrated, but also revealing some of our deepest beliefs about our sense of humanness.

As it floats in the womb, the fetus comes to develop a fantasy in which its skin is continuous with that of its mother. Over time, as the child develops its own sentient identity, separating its skin from that of its mother, it begins to see itself and the culmination of all of its inner psychological processes as existing within the skin. Our skin, the largest organ of our body and the most visible representation of us, comes to house our ego. The concept of the ego-skin, proposed by French psychoanalyst Didier Anzieu, who said "the skin is the cradle of the soul," seeks to explain why the skin becomes a battleground for psychic trauma, just as much as it is a playground for sensual pleasure.[11]

Because our sense of who we are, as both part of and separate from our world, is so embedded in our skin, it is no wonder that a breach of the skin can feel so meaningful. Some people cut themselves to give their scarred soul a face, grounding their suffering in their bodies.

"I suffer therefore I am," wrote Anzieu.[12]

Contemporary medicine is bent on bending the human body to its own tools and instruments, which have largely been hijacked by commercial interests that use drugs, devices, and procedures to monetize human fallibility. Our conventional medical approach might help someone with an intractable itch after a burn injury but fails those who continue to itch long after their scars have faded. A multidisciplinary approach to chronic itching, one that integrates both physical and psychological elements of therapy, treating every aspect of what causes us to uncontrollably scratch, is in fact the gold standard of its management.[13]

A similar approach is needed for the management of chronic pain, and it may require us to dig out some of our most ancient remedies.

After I hurt my back, I essentially became bedridden. Every move of my body, even a slight shift in balance, hurt so much I worried I would snap my spine in half. After the spine surgeon told me he thought an operation might make matters worse, he suggested I go see a physical therapist.

A friend helped me over to the sports center at our medical college campus, the same place where I had gotten injured, where the physical therapist happened to work. The jovial therapist spoke to me for a bit before putting me through the wringer. He had me on all fours, alternately arching and flexing my back. I thought I could feel every one of my vertebrae grating against its neighbors. He had me sit on the floor with my legs crossed, rotating my head, neck, and body from one side to another. This wasn't helping. My back hurt even more than before. The therapist assured me that I wouldn't split my spine open, that I wouldn't become paralyzed, scenarios that I was actively rehearsing during my contortions. He asked me to perform the set of exercises three to four times a day, but I could barely contemplate redoing them even once. I felt like I had just been assaulted.

Ancient humans exercised not because they liked to but because they had no choice. As humans evolved from monkeys, they went from being quadrupedal tree dwellers to bipedal walkers. To achieve this transition, they gave up strength for endurance, primarily because it took them so long to track and chase their prey. For bushmen in the Kalahari Desert, for example, an average endurance hunt covers nineteen miles. We had to be able to cover a distance just shy of a marathon every time we needed to put food on the table.[14]

Therefore, an even stronger human instinct than to exert is to rest whenever given the chance, to conserve the nutrition garnered from that arduous hunt, which we didn't have the option to do much of until recently, with the advent of cars and motorcycles and an automated economy reliant less on manual labor than on swipes on a touchpad.

The physician Susruta, who taught medicine at a university in Benares, India, was perhaps one of the first persons to ever write a prescription for exercise for his patients, which he did in 600 BC. He asked his patients to exercise every day, for "diseases fly from the presence of a person habituated to regular physical exercise." Yet exercise remains anathema for the injured. For someone in pain, just the thought of activity can be dreadful. This fear of movement, called kinesophobia, is present in 50 to 70 percent of folks with persistent pain. Kinesophobia is so common for a reason: if you have fractured your arm, the last thing you want to do is aggravate it by flinging it around. Yet this rational fear spills over into most chronic conditions, and in such cases immobility becomes harmful.[15]

The fear of movement is the mortar that helps erect the prison of pain. As people become deconditioned, losing strength, pliability, and balance, they become even more vulnerable to injury. By making people more watchful, kinesophobia makes them even more sensitive to pain. Longitudinal studies show that a greater degree of kinesophobia is associated with the development of more crippling disability and pain and lower quality of life, independently of other factors.[16]

For these reasons, and despite my reservations, my physical therapist was right. Exercise is a vital part of any multidisciplinary approach to managing chronic pain. Even as it can lead to initial increases in pain for some, exercise has been shown to be entirely safe for those with chronic pain and is associated with improvements in pain intensity and physical function. Exercise, actually, is known to be a most potent stimulant of the body's innate painkillers.[17]

Trapped in a boundless burrow with my life in tatters, with every inch causing a ton of torment, I knew I had to dig my way out or die trying.

In my dorm room, I finally decided to slowly pull my knees close to my chest. I couldn't get them any closer than the halfway mark though. The boggy heat left me soaking in sweat with even minimal exertion. Holding them from the back wrapped in my arms, I stayed rolled up like a ball for as long as I could. It hurt even more as I uncoiled. With surgery out of the question and my aversion to opioids, this was my only shot. So I pulled my legs up again.

The next day I walked over to the physical therapy center with an exercise prescription in hand for the receptionist. I was led into a dark room in the back with beds separated by thin linen curtains. I would become intimate with this room over the next year, returning almost every day, at times more than once. To be rid of pain, I would have to brave a whole lot of it. During this time, unable to stand in the operating room during a long operation, distracted while seeing patients in the ward by the sight of an empty chair, I was barely able to function as a medical student. It was the toughest time I had spent within my body. I had forgotten what it was like not to be in pain and began to lose hope that pain would not be the last thing I felt as I fell asleep and the first thing I woke up to.

And then, gradually, without much fanfare, I began to get better. I could bend my knees to my chest. I started exercising more frequently, often using movement as a painkiller to douse the flames before they became an inferno. Exercise saved my life.

But what if it wasn't the exercise? What if it was the people who worked with me patiently? Who attended to my concerns? Who didn't minimize or doubt how I felt but simply wanted to help? Who at times left me in the room to relax and recover even after the center closed? Who accommodated me even when I didn't have an appointment scheduled? What if it was their hearts, not their hands, that healed me?

Through five years of medical school and seven years of clinical training I have searched almost constantly for the answer to a single question: What makes a good doctor? Is a good doctor someone who helps patients live longer or feel better? How can one balance the need to be present for the patient in front of them, while also meeting the needs of the countless others waiting to be seen? How can one practice meticulous care for one's patients while also being there for one's family, tracking down the ever elusive unicorn named work-life balance? Is a good doctor judged by the degrees framed on their walls, the number of research papers they publish, or their popularity on social media?

I have come to believe a good doctor has an almost magical quality to feel what their patient feels—to see how they view the world, understand where they have been and what they have seen, all in an instant—as well as the knowledge and expertise to respond ethically to what they see from the other side. This superpower has a name: empathy. This ability to understand what another person is experiencing from their frame of reference might well be the key to ending our modern-day "paindemic."

When we see another in pain, something funny happens in our heads: the parts of our brain responsible for giving us that hurting feeling light up. Like itch, pain is contagious: our minds attempt to actually re-create in ourselves what another might be physically undergoing. How much of that pain is reimagined depends on our relationship with the person in agony. We feel more for humans over robots, friends over strangers, and lovers over friends.[18]

The care of the patient with chronic pain requires clinicians to be their most empathic. Those with chronic pain often present with no visible scars, no obvious injury. They often have seen many different clinicians without finding an answer for their woes or an end to their suffering. They not only experience injustice but come to view the very arc of existence as unjust, as if the very universe bends toward suffering. On the other hand, the physician often does not have a silver bullet: there is no antibiotic for pain, no vaccine against suffering. This sense of futility is just one reason why physicians rate 41 percent of visits with patients in chronic pain as "difficult" compared to 15 percent of average visits. Physicians are much more likely to report frustration with chronic-pain patients, consider them manipulative, hope they won't return, and feel they are self-destructive.[19]

A core mechanism through which human beings convey empathy is touch. In an experiment that looked at the painkilling effect of contact, researchers studied the brain waves of romantic partners wearing electrodes on their heads simultaneously. When two people bond, their brain waves actually imitate each other's, as touch blurs the boundaries of the self and the other. The experiment, which recruited twenty-two straight couples, showed that when the participants were subjected to pain, their lover's touch not only increased the empathetic connection between them but actually reduced how much discomfort the subject felt. This connection is deepest when one of the participants is in pain and the other is able to hold them. The weakest connection occurred when one of the partners was hurt, but the other partner was not able to touch them. Yet all touch is not equal: the closer the two people were, the more pain was relieved.[20]

What makes the touch of a loved one such a potent painkiller?

The answer might be the so-called love hormone itself, oxytocin, which is secreted by the hypothalamus in the brain. Oxytocin appears to be another way that the body finds comfort from within. Emotional connectedness and physical entanglement cause the hypothalamus to release oxytocin, as does childbirth. And while regularly given intravenously during labor, oxytocin is increasingly being used for chronic pain, although high-quality clinical evidence is still pending.

While the data are mixed, one mechanism through which oxytocin improves pain relief is its enhancement of the placebo effect, one of the body's most potent shields against suffering.

The placebo effect doesn't just reveal the deep well of neurochemical resilience contained within us to help relieve pain all by ourselves. It is also central to quantifying the impact a kind clinician can have on the distressed.

When I was in school, the night before a major examination, my mind would race, and I often needed something to tame the hordes of binomial equations and biological pathways I had crammed. Wary of sleeping pills, Mama would give me half of my asthma tablet, an antihistamine with a deep sedating effect. Therefore, for most of school and medical school, I had a benign medication I could take that would guarantee restful sleep prior to a test—that is until the day in pharmacology class when I learned that my trusted antihistamine was entirely nonsedating. The medicine did nothing. All these years I had been putting myself to sleep.

The placebo response is one of our best-studied neuropsychological phenomena. It is a powerful crank that our body can turn to ease almost any discomfort we feel: nausea, itching, anxiety, fatigue, difficulty breathing—all can be comforted by our ability to mount a placebo response. The placebo effect can even modulate the responsiveness of our immune system. Not only is pain most likely to ease in response to our expectation of relief, but it is also most readily summoned when we fear its arrival—which we call the nocebo effect.[21]

The placebo response is more powerful than most imagine. In the majority of trials, the placebo effect is as potent as the treatment being tested. On the other hand, up to a quarter of participants in clinical trials stop taking placebo pills because of side effects that they experience, the most common of which is pain. Simply being falsely told that an infusion of fentanyl has been stopped abolishes the entire analgesic effect of this formidable drug. Simply being informed that you might experience side effects from a neutral pill can markedly increase the chance that those side effects will be experienced. Simply mislabeling an antimigraine drug as a placebo can almost neutralize its benefit; in fact, the effect of a placebo drug labeled as a migraine drug was similar to that of the migraine drug when it was deceptively labeled a placebo. And the more invasive or elaborate the placebo, the more dramatic the response: some "sham" procedures can engender placebo responses that can last for years, often on par with the actual surgery they are compared to. Such sham procedures include injections of saline into the spine and joints and a knee surgery in which, although an incision was made in the skin, the knee itself wasn't manipulated.[22]

When it comes to making sense of the placebo response, we often fixate on the sugar pill or the sham procedure. Yet the most important conjurer of the placebo response is in fact the person prescribing it.

Ted Kaptchuk has no business being a professor at Harvard. A trained acupuncturist with a degree in Chinese medicine from an institute in Macao, he is one of the few, if any, faculty at Harvard Medical School with neither an MD nor a PhD. His diploma is not even recognized in Massachusetts. Yet he is one of the most innovative researchers in the world and has done more than any other scientist to help us better understand the effect doctors can have on their patients.

In 2008, Kaptchuk published the results of a groundbreaking trial. He enrolled 262 patients with irritable bowel syndrome, a chronic debilitating condition characterized by pain in the abdomen with either diarrhea or constipation. He divided the patients into three groups. In one group, patients were placed on a waiting list for future treatment, and in the second group they received "sham acupuncture" using a special device made to look exactly like a real acupuncture device. There was little difference in how patients on the waiting list and those receiving the sham acupuncture felt. In a third group, patients received what was called "augmented acupuncture"—while the sham acupuncture procedure itself was similar, everything else about the experience was different. Patients were greeted warmly, and interviewers listened to them intently, often repeating back their own words as they asked for clarifications. Importantly, they empathized with the person's ailment, for example, saying, "I can understand how difficult IBS must be for you." These acupuncturists received extensive training and coaching in improving their interactions with patients.[23]

The difference was astounding: patients receiving augmented acupuncture had more than twice the improvement in quality of life and symptom severity experienced by either the patients on the waiting list or those who received acupuncture without the attentiveness and compassion. In fact, the response rate among the augmented acupuncture group was as good as that for many drugs specifically approved for irritable bowel syndrome.

Yet even a placebo study like this carries the stain of deception. There is a sense that patients were manipulated and misinformed. While physicians often prescribe treatments they consider placebo, mostly in the form of vitamins, antibiotics, and supplements, few ever prescribe a true placebo, one made of sugar or flour. The American Medical Association's Code of Medical Ethics warns, "The use of a placebo without the patient's knowledge may undermine trust, compromise the patient-physician relationship, and result in medical harm to the patient."[24]

So Kaptchuk performed an even more audacious experiment.

This time he actually told his patients they were going to receive a placebo.

In a study of eighty-three patients with chronic low back pain, patients were randomized to either continue their usual treatment or receive medications clearly labeled "placebo." Patients were told that the placebo effect can be powerful and has been shown to alleviate pain in previous studies. The results were striking. After just three weeks, the patients who knowingly received the placebo in addition to their usual treatment reported a 30 percent reduction in disability, while those receiving their usual treatment alone had no change whatsoever. Improvements in pain were twice or more in the open-label placebo arm than in the usual-treatment arm. These improvements were noted despite 70 percent of participants initially reporting skepticism that the placebo would be helpful.[25]

It appears that placebos given without deception might be even more effective than placebos given without the patient's knowledge. A systematic review of studies assessing the effect of open-label placebos across a range of conditions—cancer-related fatigue, depression, attention deficit/hyperactivity disorder, migraines, back pain, irritable bowel syndrome, and allergic rhinitis—showed greater improvement among them compared to traditional covert placebos.[26]

The secret sauce of the placebo effect is therefore the medical ritual, one greatly enhanced by kindness. Placebos can't heal a fracture; they can't sew back a ruptured appendix. But they can make the difference between a healer and a technician. When I interviewed Ted Kaptchuk for the *Boston Globe* in 2014, he told me that any physician could maximize their personal placebo effect by "making a commitment to being present." In a study of about three thousand patients with chronic pain, those who perceived their doctors as more empathetic had a greater reduction in pain in the clinical setting.[27]

And yet empathy is not what many patients with chronic pain experience when they go for medical care. Modern healthcare and the pain medicine specialty face a reckoning, one largely spurred by a failure to provide compassionate care to those who come knocking. This is a critical reason why the dearth of female physicians in pain medicine might be so detrimental, since female physicians are generally more empathetic toward their patients than male physicians.[28]

Yet gender diversity alone may not be enough to make up for the empathy gap in pain medicine. A study of seven hundred American physicians showed that not only are female doctors more empathic than male doctors but that anesthesiology—the specialty that trains the majority of American pain physicians—has the least empathic practitioners of any medical specialty. Medical specialties are often grouped into those that are person centered and those that are technology centered, with anesthesiology firmly in the technology-centric group. Yet if ever a condition required a person-centric physician, it would have to be chronic pain.[29]

A lack of Black physicians might also be why the agony of Black Americans is less likely to be appreciated appropriately. An extensive body of research shows that we feel another's pain most acutely when we feel like they are part of our same group: when we support the same soccer team, practice the same religion, and belong to the same race.[30]

This racial intergroup bias in empathy is ubiquitous but not insurmountable. Overcoming racial bias in empathy could be central to achieving equity not just in pain management but throughout medical care and society at large. Therefore, perhaps more important than discovering a new drug for pain relief is training physicians to be kinder, redesigning healthcare to allow them more time with their patients, and rejiggering incentives away from doing more and toward doing the right thing.

There is ample evidence that empathy for others' pain can be actively developed. Simply living together and interacting with people of other races is associated with increases in empathy.[31]

Another key factor that mediates implicit bias is that people emphasize the differences between races rather than what unites them. A process called individuation training nudges people to focus on other people as individuals with unique traits and characteristics rather than as members of a particular group. In one experiment, focusing on how much pain people from other races were feeling rather than the color of their skin increased the empathetic response generated, eliminating intergroup bias. The same was true when individuals were made to believe that the member of the disfavored racial group was now part of their team.[32]

Empathy is the key to helping people with chronic pain and to overcoming the biases so prevalent in how people are treated. Yet some are nihilistic about whether individual clinicians can be trained to be kinder. They believe empathy is like the ability to roll one's tongue: some have it, and others simply don't.

I believe that while certain individuals have a greater instinct for kindness that no amount of communication skills workshops can engender, a significant degree of empathy can certainly be instilled. I also believe experiences can change how one inhabits another's world. My back injury derailed my life but sensitized me to the agony of my patients, many of whom carry invisible scars like mine. Ideally, we should inculcate in clinicians the ability to relate to those in agony, without having them go through a medical calamity themselves.

But most importantly, we should operate a healthcare system that favors thoughtful and humane care through how it is structured and how the incentives are aligned. Unfortunately, though, the very opposite has occurred, particularly in American healthcare. And no one has suffered more for it than the person in pain.

John Bonica, whom *Time Magazine* called "pain relief's founding father," had a very clear vision for what he considered the best way to help those who hurt. To Bonica, the provision of comfort required the collaboration of multiple disciplines focused not just on the body or the mind but on the confluence of both: the whole person.

Speaking about the poor care that people with chronic pain

received, punctuated more by the complications rather than the effectiveness of medical interventions, the iatrogenesis that vexed Ivan Illich, Bonica wrote in 1976, "Many of these [patients] are exposed to a high risk of iatrogenic complication from improper therapy, including narcotic intoxication and multiple, often useless, and at times mutilating operations; a significant number give up medical care and consult quacks who not only deplete the patient's financial resources but often do harm; some patients with severe intractable pain become so desperate as to commit suicide." In particular, he lamented "the progressive trend towards specialization, which is conducive to each specialist viewing pain in a very narrow, tubular fashion." His solution for providing whole-person pain care was the multidisciplinary pain clinic, which took off across the United States.[33]

Multidisciplinary pain treatment respects the irreducible complexities of the experience of pain. When Bonica, an anesthesiologist by training, set up the first multidisciplinary pain clinic in the world in 1961, it included surgeons as well as psychologists, psychiatrists, and physical therapists. The evidence base for such an approach is very strong, and unlike other approaches that overwhelmingly focus on drugs or procedures, there are few, if any, risks involved.[34]

When Carol Warfield opened her multidisciplinary pain clinic in 1979, as she recalls, it was a different world. "In the 1980s or 1990s, you never prescribed an opioid for chronic pain," she told me. For many cases, experts from around Harvard would get together to come up with a strategy to help individual patients. With the gift of time, the members of the team could converge on the best path forward.

And then everything changed. Just as quickly as multidisciplinary care flourished, clinics started to get shuttered across the United States. Compared to more than one thousand in 1999, in 2021 there were only fifty-one interdisciplinary pain rehabilitation programs accredited by the Commission on Accreditation of Rehabilitation Facilities in the country; with eighteen in Texas alone, few in the rest of the country have access to them.[35]

Today how we manage pain is the same as, if not worse than, what Bonica described way back in 1976. How did that come to be?

A major factor in the decline of multidisciplinary care was insurance companies. Many insurance companies don't cover multidisciplinary care because they do not view it as cost-effective. Ironically, this strategy has often led to patients receiving either ineffective procedures, which insurance companies pay a much greater amount for, or opioids, which require much less paperwork and provide some relief in the short term but fail to help in the long run. Even when approved, coverage for multidisciplinary care is often delayed through the imposition of administrative burdens on both clinicians and patients. These delays often worsen patients' symptoms, and many simply give up.[36]

Yet an even greater reason why multidisciplinary care has all but expired is the rampant corporatization of medicine.

American healthcare is established on a fee-for-service model. So essentially a doctor visit is no different from a trip to the shopping mall: the more you put in your cart, the more you pay. The difference is that in healthcare, the real shopper is the doctor. Patients don't know what options exist, how much they might cost, and what their risks or benefits might be. The patient is therefore captive to the clinician's preference. This creates a conflict of interest, because the more expensive the items they charge for, the more doctors and hospitals get paid.

Not long ago, physicians rarely talked about what Carol Warfield calls the "M word." "We never talked about money or how much a procedure paid. We just did what was best for the patient," she said. That changed as the desire for reimbursement overwhelmed medicine's moral prerogative. "Physicians in pain medicine went towards procedures because they were more lucrative."

In a healthcare environment with constant pressure by higher-ups to generate more revenue, even the most ethical physicians might find themselves thinking more often in terms of what is good for business.

While individual physicians are trained to abide by a strict moral code, the institutions and systems in which they work are not obliged to operate within those parameters. Yet physicians bear considerable responsibility for buckling to these pressures.

"Interventional pain management specialists may take advantage of vulnerable patients with unremitting pain who perform highly reimbursed procedures with little evidence supporting their effectiveness," writes Allen Lebovits, a pain psychologist, in the journal *Pain Medicine*. "Listening to the patient and observing his/her pain behavior can take time, with 'time' becoming a commodity for which the 'business' of pain medicine often does not allow."[37]

A healthcare system that prioritizes health over profit can transform how the person in extremis is treated. What the treatment of pain might look like in a less ravenous healthcare environment is far from hypothetical, for such an example exists right here in the United States: the Veterans Affairs Health System (VA).

The VA, the largest integrated healthcare system in the United States, is bucking every turn for the worse taken by American medicine. Even as opioid prescription remains dangerously high for civilians, the VA has achieved a 40 percent reduction. As multidisciplinary care dwindles in the United States, it has become the standard of care in the VA system, with almost all VA facilities offering complementary and integrative health options for pain, ranging from yoga to meditation to hypnosis, in addition to the standard medical and surgical options.[38]

The VA's approach to management is important for many reasons. Unlike conventional medical centers, the VA does not operate on a fee-for-service model. A global budget is provided to centers, leaving them to decide how to use resources to achieve the best outcomes for patients. One important reason I chose to work for the VA after I finished my clinical training was that I felt the incentives for both the patient and the healthcare system are fully aligned in such an environment: what is best for the patient is what is best for the system.

The VA is also funding major clinical trials in chronic pain, such as the Strategies for Prescribing Analgesics Comparative Effectiveness (SPACE) trial, that are providing much-needed insights into the cavernous unknowns of pain.

Yet, for multidisciplinary pain management to truly replace the abject manner in which we provide pain relief today, we need buy-in from the most important stakeholder of them all: the person in unremitting anguish. Having been oversold the promise of cures through drugs or surgeries, many are hesitant to consent to one of the central pillars of multidisciplinary pain care: acceptance.

To step into a future in which every patient receives an individualized approach to help them live with pain, we will have to reimagine the body's fundamental relationship with how it hurts.

Recently, after a hiatus of many years, I went to the dentist. My last trip had been such a bloodbath, I just didn't have the stomach to return. But at some point the aching in my mouth became too hard to ignore.

The picture the dentist painted was not pretty: My gums were so inflamed that even the slightest prod caused blood to ooze. They were also exquisitely tender. I had wisdom teeth popping out at jagged angles, crowding my jaw. She warned that the cleanings would be intensely painful and that I would require three separate appointments. These appointments are usually performed months apart, but my work schedule only permitted me to have them performed on three consecutive days.

The first cleaning offered a relentless assault of spine-tingling torment. The high-pitched shrieking of the scaler and the taste of blood in my mouth set a macabre background score for the stabbing and scraping of my exsanguinating gums. I tried to force my body to relax, but I would respond to any new movement of the instrument by stiffening like a slab of concrete. I tried to focus on my toes or imagine myself playing video games (my version of relaxing by the beach). Neither worked. The office was cold, but I was drenched in sweat from head to toe. My body was flushed with swirling stress hormones long after I returned from the clinic.

When I went in the next day, I expected things only to be worse, given how battle weary my mouth already was. But halfway through the procedure, the dentist remarked about how much calmer I looked. I felt even better on my third trip. On reflection, I think I know why. I wasn't in any less pain, but with every subsequent trip, I knew what to expect. It made all the difference.

Pain—whether it burns, aches, stabs, tingles, or crushes—says the same thing no matter what language it speaks: it articulates a fundamental disagreement between where the body wants to be and where it is. It is a statement that *demands* action. This plays right into the hands of a healthcare industry increasingly concerned about making money.

A trial of patients with diabetic neuropathy, a condition that causes tingling and discomfort in the hands and knees of people with long-standing diabetes, illuminates the key to living with pain. Patients were randomized to receive usual care or a cognitive therapy program focused on training them to acknowledge the disagreement in their body and to help them understand the power of a choice that they might have to make: whether to live their lives despite the pain or to curb their lives in order to either avoid or control the pain. This approach is called acceptance and commitment therapy (ACT), which represents the most modern wave of cognitive behavioral therapy, which in turn is the most common and best-studied form of psychotherapy.[39]

In this trial, patients who received this therapy experienced a great improvement in both pain perception and pain acceptance. Pain acceptance, measured using validated questionnaires, reflects the bearer's willingness to let go of attempting to control the pain and to experience it in the course of an active life driven by their values.

Pain acceptance is controversial, since many patients incorrectly equate it with resignation to the inevitability of pain, but the evidence of its effectiveness could not be more striking. Multiple trials have shown that even when provided via the internet, ACT can be very effective at reducing how much people suffer and expanding how much of their lives they want to live. ACT is designed to trade a person's desire to avoid unwanted thoughts, emotions, and situations for a psychological flexibility enabled through openness to engage in activities by defusing the self from its sensations.[40]

ACT can greatly augment other treatments we have for those with persistent pain. In one clinical trial in the United Kingdom, physical therapists were trained to help patients prioritize functioning over pain reduction and to use acceptance, mindfulness, and values-based action. A special aspect of the trial was that patients themselves were intimately involved in its design; one was even a coapplicant on the grant that funded the trial. The trial randomized 248 people with chronic low back pain to receive either physical therapy informed by ACT or conventional physical therapy. Even though the physical therapists using ACT were not psychologists and ended up spending less time with the patients than those providing conventional physical therapy, the results were quite impressive.[41]

While patients receiving the ACT-enhanced physical therapy did not have less pain, they had greater improvement in disability, pain-related interference, physical function, and work and social adjustment. Importantly, no patient experienced an adverse event from ACT despite a greater emphasis on working through the pain. In fact, patients rated the ACT-enhanced physical therapy as more effective than usual physical therapy.

Acceptance can come across as a bitter pill for patients who have already endured a protracted struggle for relief and legitimacy. For many, their interaction with the healthcare system can be as cold and sharp as a metal scalpel.

It is fully understandable why many patients might reject such an approach. Anna Hamilton, a disability advocate, wrote in a blog, "'Pain acceptance' may be useful for some patients, and I do think that some degree of accepting one's pain is necessary—

but the one-size-fits-all approach to 'just accept the pain,' especially if that advice is not part of a multi-faceted care plan, could be very dangerous."[42]

Patients worry that acceptance might be dangerous for it might abruptly end their voyage toward an elusive diagnosis or an untried intervention that could turn their lives around. They may hear acceptance as submission, a sign of abandonment, a dereliction of medicine's duty.

Asking patients to accept their pain might make them feel like they are not being given a choice. On the contrary, pain acceptance is about helping patients make the most important choice in their journey: choosing a life dominated by pain or one dominated by agency.

If ACT is done right, with kindness and empathy, many patients come to find such a way of life helpful. In the UK study, of 478 patients screened for eligibility for enrollment, a whopping 85 percent entered the trial, a very large number for such a research study. Of those who were enrolled in the study, an impressive 88 percent completed the study fully and provided follow-up data.[43]

Armed with this knowledge and a strong belief in ACT, when I met the patient, mentioned in Chapter 1, who was wasting away from the pain of shingles and had been admitted to the hospital, I sat on the bed next to him. He had already undergone a million-dollar workup and received countless drugs and procedures, all to no avail. Though I was skeptical we would find some new cause for his pain or uncover a treatment that would rid him of his agony, I reassured him that we, as a team, were fully committed to figuring out why he hurt so much and what we could do to help. At the same time, though, I confessed that the likelihood that we could make the pain go away was low.

"I want to help you live with this pain," I told him. "I want you back to playing golf again."

Perhaps because he believed that we were all in, that we would do whatever it might take to help him, his entire outlook changed. Without us making any changes to his drugs, within a few days he began to feel much better. He refused to go to a rehabilitation facility, motivated to get stronger on his own accord at home. Before he left, I asked if he was interested in seeing a pain psychologist. We were not raising the white flag, I told him, but gathering all the options we had at our disposal to close the rift that had opened up in his body. He eagerly accepted the referral.

Pain acceptance has only recently become part of the biomedical approach to helping people live with pain. But even a cursory review reveals that it has been one of the primary ways we have lived with pain throughout our history. Buddhists often subdue intolerable pain by confronting it fully, not with boisterous bravado but with quiet reflection or meditation. Most cultures have encouraged cohabitation with pain, emphasizing peaceful coexistence rather than confrontation.

The most cutting-edge pain science is not just revealing new things about pain but reintroducing a way to inhabit our body that a burgeoning corporate machine selling visions of immortality and painlessness has compelled us to forget. The central premise of corporate healthcare is to subdue the individual, indoctrinating us with the notion that the remedy for our maladies can only be prescribed by a physician's pen or carved by a surgeon's blade.

To course-correct our approach to pain, we need to change the story of chronic pain—pushing back on the voices attempting to convince us all pain is catastrophic and life threatening and needs immediate attention over everything else right now. Because even as we search for the answers all over the world, in everything from ancient herbs to modern psychedelics, invasive procedures to contortionist postures, the ultimate relief for chronic pain has been lying deep within us this whole time.

A pain-free world can never be achieved—and, one might argue, *should* never be achieved. Our quest to successfully prolong life has also unleashed unforeseeable suffering. Our efforts to quell the epidemic of pain have birthed a demon that takes even more lives than the poison for which it was a supposed antidote. We have a choice to make. Will we take the same road we have been on, in pursuit of a vision of pain corrupted by the influence of corporate machinators? Or will we own pain and reimagine it, as we always have, as an experience that we live with and as an emotion that we feel rather than simply as a physical sensation we perceive?

Over the course of a life lived with pain, a question took root in my head that I found increasingly difficult to shake off: Was all my pain, and all the pain that I had ever felt, all just in my head? Was it a figment of my overzealous imagination, a manifestation of my overvivid recollection?

For the person in pain, "it's all in your head" is one of the most dismissive, unhelpful things they can hear. Pain is central to the daily narrative of so many, and questioning its reality is essentially erasing the existence of the person enduring it.

Yet it's difficult, given what we know about the origin of pain in our bodies, to come to any other conclusion. Of all our sensations, pain is the most compliant. Of all that we feel, pain is the sensation most of the mind rather than the body. The body detects nociception, but it is the mind that generates the experience of pain, the mind that causes us to suffer.

All pain, in fact, is in our heads.

This shouldn't be a controversial statement. But the pervasive domination of mind-body dualism in medicine renders it so.

During the Renaissance, mind-body dualism allowed religious forces to keep dominion over the supposedly sacred mind, while biology and medicine took hold of the body, allowing science a clean break from theological control. Medical practice was divided between doctors of the mind, later called psychiatrists, and doctors of the body. And as doctors of the body found tremendous success, our understanding of the mind remained stunted. After an initial rush toward mass institutionalization and tranquilizers, even more recent psychiatric developments like antidepressants have barely been shown to work. And as the attention paid to physical ailments has increased, the stigma of mental illnesses has remained pervasive.[44]

If nothing else, pain presents the strongest argument against the fallacy of mind-body dualism. The mind, our entire sense of consciousness, is created by the brain and the body. We know too well that what affects the body affects the so-called mind—such that under most circumstances nociception causes pain, which leads to suffering. But pain makes clear that what upsets the so-called mind can change what the body feels, reversing the chain of events from a signal that travels from the bottom up to one that descends from top to bottom.

Our nerves don't know where the mind ends and the body begins. Perhaps there is no clearer example of their total enmeshment than that provided by the growing field of epigenetics.

Epigenetics refers to how gene expression can be modified by major life events. Just growing up in a stressful environment, for example, can lead to chemical changes in our genes that affect how they function. To see just how much of an influence our lifestyles and environments can have on how tender our nerves are, researchers in the United Kingdom studied twenty-five pairs of identical twins to look for differences in their sensitivity to heat. While each set was born with the same genes, the researchers found clear differences in when some twins found a heat probe had become too hot. The researchers concluded that due to epigenetic changes stemming from different life events, the expression of a gene associated with pain— TRPA1—varied between the twins, appearing to explain their differing sensitivity to heat pain.[45]

Our subjective experiences can fundamentally change our physiology. People who have had adverse childhood experiences likely undergo epigenetic modifications that make them highly prone to developing chronic pain later in their lives. Such changes might be one reason people who experience racial discrimination have been found to have a greater sensitivity to discomfort.

Modern medicine is slowly beginning to understand that the human mind and body are not distinct entities but a representation of the same underlying biological processes. Our consciousness is the sum product of our cognitive processes, and everything we feel is its outcome. Hunger is a feeling produced not by an empty stomach but by our minds. Pleasure is something we feel not in our skin but rather in our brains. Pleasure can be felt without the skin's being caressed, and you can crave ice cream even on a full stomach.

The most important thing doctors and nurses and physical therapists can do is center their practice in empathy and kindness. But to allow kindness to become the standard of care, our medical schools and training programs have to make it a point of emphasis, and our health system has to evolve. We need to take a multidisciplinary approach to pain that provides patients with all the tools we have to diminish how much they hurt. We need to make person-centered care a reality rather than a buzzword by shifting the way providers are paid to reflect how patients do, rather than what the system does to them. The reward for designing a healthcare system that provides care and love to all might resonate beyond the walls of hospitals and clinics— it will be the keystone for a more just and equitable society.

Instead of using "all in the head" to belittle, we should use it to empower. Some of the most promising strategies for pain relief amplify our body's own mechanisms for achieving comfort. Instead of having people look to external sources for respite, medicine needs to help people maximize their own inner ability to achieve harmony. Instead of waging a war on pain, trying to conquer it at all costs, we must rediscover a new way forward. Our incessant attempts to control how we hurt, to bend it to our will, have made the discord within worse, as pain has come to dominate us in return. The best way forward will be to reconnect with a way of being that is innately our own, a state of actuality that connects us to our past, anchors us in the present, and provides hope for the future: living well despite the pain.

ACKNOWLEDGMENTS

The day in 2008 when I ruptured my back, everything changed, as I then unknowingly took my first excruciating step toward writing this book. That I am able to write about the worst of my pain in the past tense, though, is nothing short of a miracle, one that few are ever lucky enough to witness. To right all that went wrong, it took an army of good Samaritans: the attending physicians who let me leave early from clinic or the operating room, the physical therapists who at times helped me free of charge, and the friends who stuck around after so many others left me behind. But I am alive and well to tell this tale because of one person: the girl I was, and remain, in love with, my now wife, Rabail. For all the ways I have replayed the agony in my back, what explains my recovery has nothing to do with anything one can find in a textbook or in a research paper. Her love was my balm and her care the glue that put my broken body and soul back together. Yet the specter of chronic pain has never truly left my life, always threatening to return with one careless workout or one unforeseen accident, and I have been forever changed as a person for it.

The Song of Our Scars is ultimately about pain and the people who live with it, who often go unheard, their suffering often delegitimized by society and the healthcare system. To write this book, I spoke with many people living with pain or caring for someone who hurts. They opened their lives to me, and for that I am eternally grateful. I hope that the candor of those I spoke with can help readers suffering in silence. I do foresee many in unremitting agony disagreeing with portions of the book, but I hope they also see this work coming from a place of kindness and intellectual rigor. While we all suffer in our own ways, one purpose of writing this book is to show how much pain connects us, not just with each other but with the entire kingdom of living beings. This is not the first word on pain and certainly not the last.

Working as a physician, as essential to me in writing this book as my own lived experience with pain, has been the greatest privilege of my life. I went to medical school at the Aga Khan University in Karachi, Pakistan, and trained in internal medicine at the Beth Israel Deaconess Medical Center in Boston, Massachusetts, and in cardiology at the Duke University Medical Center in Durham, North Carolina, and I could not have asked for a better education in science, medicine, and humanity. I would also like to thank my colleagues at the VA Boston Healthcare System, Brigham and Women's Hospital, and Harvard Medical School—I am honored to practice medicine, conduct research, and teach the very best at these hallowed institutions.

Finally, a stellar team transformed this work from an idea into a manuscript. My agent, Don Fehr, told me the day I first wrote an essay on chronic pain for the *New York Times* that there was a book here. And he was right. He helped me find an excellent editorial team at Basic Books led by Eric Henney, with whom I connected the very first time we spoke on the phone. He pushed me to be better and guided this book to where it needed to go. Emma Berry took the baton and helped this work cross the finish line.

NOTES

INTRODUCTION

1   Hoffman KM, Trawalter S, Axt JR, Oliver MN. Racial bias in pain assessment and treatment recommendations, and false beliefs about biological differences between blacks and whites. *Proc Natl Acad Sci USA.* 2016;113(16):4296–4301.

2   James SL, Abate D, Abate KH, et al. Global, regional, and national incidence, prevalence, and years lived with disability for 354 diseases and injuries for 195 countries and territories, 1990–2017: a systematic analysis for the Global Burden of Disease Study 2017. *Lancet.* 2018;392:1789–1858.

3   Goldberg DS, McGee SJ. Pain as a global public health priority. *BMC Public Health.* 2011;11:770; Blanchflower F, Oswald, A. Unhappiness and pain in modern America: a review essay, and further evidence, on Carol Graham's Happiness for All? NBER Working Paper No. 24087. 2017; Zelaya CE, Dahlhamer JM, Lucas JW, Connor EM. Chronic pain and high-impact chronic pain among U.S. adults, 2019. *NCHS Data Brief.* 2020;390:1–8; Institute of Medicine. *Relieving Pain in America: A Blueprint for Transforming Prevention, Care, Education, and Research.* Washington, DC: National Academies Press; 2011; Craig KD, Holmes C, Hudspith M, et al. Pain in persons who are marginalized by social conditions. *Pain.* 2020;161:261–265; Mills SEE, Nicolson KP, Smith BH. Chronic pain: a review of its epidemiology and associated factors in population-based studies. *Brit J Anaesth.* 2019;123:e273–e283.

4   Wiech K. Deconstructing the sensation of pain: the influence of cognitive processes on pain perception. *Science.* 2016;354:584–587.

CHAPTER 1. THE INTERPRETATION OF AGONY

1   Moehring F, Halder P, Seal RP, Stucky CL. Uncovering the cells and circuits of touch in normal and pathological settings. *Neuron*. 2018;100:349–360.

2   Cahill JF, Jr., Castelli JP, Casper BB. Separate effects of human visitation and touch on plant growth and herbivory in an old-field community. *Am J Bot*. 2002;89:1401–1409; Xu Y, Berkowitz O, Narsai R, et al. Mitochondrial function modulates touch signalling in *Arabidopsis thaliana*. *Plant J*. 2019;97:623–645.

3   Trewavas A. Aspects of plant intelligence. *Ann Bot*. 2003;92:1–20.

4   De Luccia TP. *Mimosa pudica, Dionaea muscipula* and anesthetics. *Plant Signal Behav*. 2012;7:1163–1167; Murthy SE, Dubin AE, Whitwam T, et al. OSCA/TMEM63 are an evolutionarily conserved family of mechanically activated ion channels. *Elife*. 2018;7; Gremiaux A, Yokawa K, Mancuso S, Baluska F. Plant anesthesia supports similarities between animals and plants: Claude Bernard's forgotten studies. *Plant Signal Behav*. 2014;9:e27886; Dillard MM. Ethylene—the new general anesthetic. *J Natl Med Assoc*. 1930;22:10–11.

5   Bastuji H, Frot M, Perchet C, Hagiwara K, Garcia-Larrea L. Convergence of sensory and limbic noxious input into the anterior insula and the emergence of pain from nociception. *Sci Rep*. 2018;8:13360.

6   Pain terms: a list with definitions and notes on usage. Recommended by the IASP Subcommittee on Taxonomy. *Pain*. 1979;6:249.

7   Raja SN, Carr DB, Cohen M, et al. The revised International Association for the Study of Pain definition of pain: concepts, challenges, and compromises. *Pain*. 2020;161:1976–1982.

8   Sneddon LU. Evolution of nociception and pain: evidence from fish models. *Philos Trans R Soc Lond B Biol Sci*. 2019;374:20190290; Batista FLA, Lima LMG, Abrante IA, et al. Antinociceptive activity of ethanolic extract of *Azadirachta indica* A. Juss (Neem, Meliaceae) fruit through opioid, glutamatergic and acid-sensitive ion pathways in adult zebrafish (*Danio rerio*). *Biomed Pharmacother*. 2018;108:408–416.

9   Rose JD, Arlinghaus, R, Cooke, SJ, et al. Can fish really feel pain? *Fish Fish*. 2014;15:97–133.

10 Dunlop R, Millsopp S, Laming P. Avoidance learning in goldfish (*Carassius auratus*) and trout (*Oncorhynchus mykiss*) and implications for pain perception. *Appl Anim Behav Sci*. 2006;97:255–271.

11   Millsopp S, Laming P. Trade-offs between feeding and shock avoidance in goldfish (*Carassius auratus*). *Appl Anim Behav Sci*. 2008;113:247–254.

12   Dunlop R, Millsopp S, Laming P. Avoidance learning in goldfish (*Carassius auratus*) and trout (*Oncorhynchus mykiss*) and implications for pain perception. *Appl Anim Behav Sci*. 2006;97:255–271.

13   Beecher HK. Relationship of significance of wound to pain experienced. *J Am Med Assoc*. 1956;161:1609–1613.

14   Ashley PJ, Sneddon LU, McCrohan CR. Nociception in fish: stimulus-response properties of receptors on the head of trout *Oncorhynchus mykiss*. *Brain Res*. 2007;1166:47–54.

15   Langford DJ, Crager SE, Shehzad Z, et al. Social modulation of pain as evidence for empathy in mice. *Science*. 2006;312:1967–1970.

16   Gioiosa L, Chiarotti F, Alleva E, Laviola G. A trouble shared is a trouble halved: social context and status affect pain in mouse dyads. *PLoS One*. 2009;4:e4143; Dunlop R, Millsopp S, Laming P. Avoidance learning in goldfish (*Carassius auratus*) and trout (*Oncorhynchus mykiss*) and implications for pain perception. *Appl Anim Behav Sci*. 2006;97:255–271.

17   Yong MH, Ruffman T. Emotional contagion: dogs and humans show a similar physiological response to human infant crying. *Behav Process*. 2014;108:155–165.

18   Guesgen MJ. The social function of pain-related behaviour and novel techniques for the assessment of pain in lambs: a thesis presented in partial fulfilment of the requirements for the degree of doctor of philosophy in zoology at Massey University, Turitea campus, Manawatu, New Zealand [doctoral]. Massey University; 2015.

19   Vigil JM, Strenth CR, Mueller AA, et al. The curse of curves: sex differences in the associations between body shape and pain expression. *Hum Nat*. 2015;26:235–254.

20   Fairbrother N, Barr RG, Chen M, et al. Prepartum and postpartum mothers' and fathers' unwanted, intrusive thoughts in response to infant crying. *Behav Cogn Psychother*. 2019;47:129–147; Brandon S. Child abusive head trauma on the rise during COVID-19. Loma Linda University Health. January 22, 2021. https://news.llu.edu/health-wellness/child-abusive-head-trauma-rise-during-covid-19. Accessed May 7, 2021.

21   Keen S. The lost voices of the gods. *Psychology Today*. 1977;11:58–64.

22   Cassel EJ. The nature of suffering and the goals of medicine. *N Engl J Med*. 1982;306:639–645.

CHAPTER 2. HOW WE HURT

1   Abdo H, Calvo-Enrique L, Lopez JM, et al. Specialized cutaneous Schwann cells initiate pain sensation. *Science*. 2019;365:695–699.

2   Yen CT, Lu PL. Thalamus and pain. *Acta Anaesthesiol Taiwan*. 2013;51:73–80.

3   Bugiardini R, Ricci B, Cenko E, et al. Delayed care and mortality among women and men with myocardial infarction. *J Am Heart Assoc*. 2017;6(8): e005968.

4   Rainville P, Duncan GH, Price DD, Carrier B, Bushnell MC. Pain affect encoded in human anterior cingulate but not somatosensory cortex. *Science*. 1997;277:968–971.

5   Rainville P, Duncan GH, Price DD, Carrier B, Bushnell MC. Pain affect encoded in human anterior cingulate but not somatosensory cortex. *Science*. 1997;277:968–971.

6   Pyszczynski T, Solomon S, Greenberg J. Chapter one—thirty years of terror management theory: from genesis to revelation. In: Olson JM, Zanna MP, eds. *Adv Exp Soc Psychol*. San Diego, CA: Academic Press; 2015:1–70; Burger O, Baudisch A, Vaupel JW. Human mortality improvement in evolutionary context. *Proc Natl Acad Sci USA*. 2012;109(44):18210–18214.

7   Yong E. Meet the woman without fear. *Discover Magazine*. December 16, 2010. https://www.discovermagazine.com/mind/meet-the-woman-without -fear.

8   Veinante P, Yalcin I, Barrot M. The amygdala between sensation and affect: a role in pain. *J Mol Psychiatry*. 2013;1:9.

9   Corder G, Ahanonu B, Grewe BF, Wang D, Schnitzer MJ, Scherrer G. An amygdalar neural ensemble that encodes the unpleasantness of pain. *Science*. 2019;363:276–281.

10 Gandhi W, Rosenek NR, Harrison R, Salomons TV. Functional connectivity of the amygdala is linked to individual differences in emotional pain facilitation. *Pain*. 2020;161:300–307.

11 Twilley N. The neuroscience of pain. *New Yorker*. June 25, 2018. https:// www.newyorker.com/magazine/2018/07/02/the-neuroscience-of-pain.

12 Mazzola L, Isnard J, Peyron R, Mauguiere F. Stimulation of the human cortex and the experience of pain: Wilder Penfield's observations revisited. *Brain*. 2012;135:631–640.

13 Isnard J, Magnin M, Jung J, Mauguiere F, Garcia-Larrea L. Does the insula tell our brain that we are in pain? *Pain*. 2011;152:946–951.

14 Mazzola L, Isnard J, Peyron R, Mauguiere F. Stimulation of the human cortex and the experience of pain: Wilder Penfield's observations revisited. *Brain.* 2012;135:631–640.

15 Benarroch EE. Insular cortex: functional complexity and clinical correlations. *Neurology.* 2019;93:932–938.

16 Bastuji H, Frot M, Perchet C, Hagiwara K, Garcia-Larrea L. Convergence of sensory and limbic noxious input into the anterior insula and the emergence of pain from nociception. *Sci Rep.* 2018;8:13360.

17 Northoff G. Psychoanalysis and the brain — why did Freud abandon neuroscience? *Front Psychol.* 2012;3:71.

18 Yong E. Meet the woman without fear. *Discover Magazine.* December 16, 2010. https://www.discovermagazine.com/mind/meet-the-woman-without -fear; Berret E, Kintscher M, Palchaudhuri S, et al. Insular cortex processes aversive somatosensory information and is crucial for threat learning. *Science.* 2019;364(6443):eaaw0474.

19 Logothetis NK. What we can do and what we cannot do with fMRI. *Nature.* 2008;453:869–878.

20 Wager TD, Atlas LY, Lindquist MA, Roy M, Woo CW, Kross E. An fMRI-based neurologic signature of physical pain. *N Engl J Med.* 2013;368: 1388–1397.

CHAPTER 3. NO END IN SIGHT

1 Benedetti C, Chapman CR. John J. Bonica. A biography. *Minerva Anestesiol.* 2005;71:391–396.

2 Loeser JD. John J. Bonica 1917–1994. Emma B. Bonica 1915–1994. *Pain.* 1994;59:1–3.

3 Bonica JJ. *The Management of Pain.* Philadelphia, PA: Lea and Febirger; 1953; Raffaeli W, Arnaudo E. Pain as a disease: an overview. *J Pain Res.* 2017;10:2003–2008.

4 Loeser JD. John J. Bonica 1917–1994. Emma B. Bonica 1915–1994. *Pain.* 1994;59:1–3.

5 Vos T, Flaxman AD, Naghavi M, et al. Years lived with disability (YLDs) for 1160 sequelae of 289 diseases and injuries 1990–2010: a systematic analysis for the Global Burden of Disease Study 2010. *Lancet.* 2012;380: 2163–2196; Zelaya CE, Dahlhamer JM, Lucas JW, Connor EM. Chronic pain and high-impact chronic pain among U.S. adults, 2019. *NCHS Data Brief.* 2020;390:1–8; Institute of Medicine. *Relieving Pain in*

America: A Blueprint for Transforming Prevention, Care, Education, and Research. Washington, DC: National Academies Press; 2011; Nahin RL, Sayer B, Stussman BJ, Feinberg TM. Eighteen-year trends in the prevalence of, and health care use for, noncancer pain in the United States: data from the Medical Expenditure Panel Survey. *J Pain*. 2019;20:796–809.

6 Mills SEE, Nicolson KP, Smith BH. Chronic pain: a review of its epidemiology and associated factors in population-based studies. *Brit J Anaesth*. 2019;123:e273–e283.

7 Blanchflower F, Oswald A. Unhappiness and pain in modern America: a review essay, and further evidence, on Carol Graham's Happiness for All? NBER Working Paper No. 24087. 2017; Sa KN, Moreira L, Baptista AF, et al. Prevalence of chronic pain in developing countries: systematic review and meta-analysis. *Pain Rep*. 2019;4:e779.

8 Mokdad AH, Ballestros K, Echko M, et al. The state of US health, 1990–2016: burden of diseases, injuries, and risk factors among US states. *JAMA*. 2018;319:1444–1472; James SL, Abate D, Abate KH, et al. Global, regional, and national incidence, prevalence, and years lived with disability for 354 diseases and injuries for 195 countries and territories, 1990–2017: a systematic analysis for the Global Burden of Disease Study 2017. *Lancet*. 2018;392:1789–1858.

9 Mokdad AH, Ballestros K, Echko M, et al. The state of US health, 1990–2016: burden of diseases, injuries, and risk factors among US states. *JAMA*. 2018;319:1444–1472; Stovner LJ, Nichols E, Steiner TJ, et al. Global, regional, and national burden of migraine and tension-type headache, 1990–2016: a systematic analysis for the Global Burden of Disease Study 2016. *Lancet Neurol*. 2018;17:954–976; James SL, Abate D, Abate KH, et al. Global, regional, and national incidence, prevalence, and years lived with disability for 354 diseases and injuries for 195 countries and territories, 1990–2017: a systematic analysis for the Global Burden of Disease Study 2017. *Lancet*. 2018;392:1789–1858.

10 Dieleman JL, Cao J, Chapin A, et al. US health care spending by payer and health condition, 1996–2016. *JAMA*. 2020;323:863–884.

11 Toye F, Seers K, Allcock N, et al. *A Meta-ethnography of Patients' Experience of Chronic Non-malignant Musculoskeletal Pain*. Southampton, UK: NIHR Journals Library; 2013.

12 Birk LB. Erasure of the credible subject: an autoethnographic account of chronic pain. *Cultural Studies ↔ Critical Methodologies*. 2013;13:390–399.

13   Rhoades DR, McFarland KF, Finch WH, Johnson AO. Speaking and interruptions during primary care office visits. *Fam Med.* 2001;33:528–532; Birk LB. Erasure of the credible subject: an autoethnographic account of chronic pain. *Cultural Studies ↔ Critical Methodologies.* 2013;13:390–399.

14   Robles TF. Marital quality and health: implications for marriage in the 21(st) century. *Curr Dir Psychol Sci.* 2014;23:427–432; Gortmaker SL, Must A, Perrin JM, Sobol AM, Dietz WH. Social and economic consequences of overweight in adolescence and young adulthood. *N Engl J Med.* 1993;329:1008–1012.

15   Junghaenel DU, Schneider S, Broderick JE. Partners' overestimation of patients' pain severity: relationships with partners' interpersonal responses. *Pain Med.* 2018;19:1772–1781.

16   Hayes J, Chapman P, Young LJ, Rittman M. The prevalence of injury for stroke caregivers and associated risk factors. *Top Stroke Rehabil.* 2009;16:300–307; Hartke RJ, King RB, Heinemann AW, Semik P. Accidents in older caregivers of persons surviving stroke and their relation to caregiver stress. *Rehabil Psychol.* 2006;51:150–156.

17   Karraker A, Latham K. In sickness and in health? Physical illness as a risk factor for marital dissolution in later life. *J Health Soc Behav.* 2015;56:420–435; AARP. Caregiving in the United States, 2020. National Alliance of Caregiving. https://www.caregiving.org/caregiving-in-the-us-2020.

18   Monin JK, Schulz R, Martire LM, Jennings JR, Lingler JH, Greenberg MS. Spouses' cardiovascular reactivity to their partners' suffering. *J Gerontol B Psychol Sci Soc Sci.* 2010;65B:195–201.

19   Spanos NP, Radtke-Bodorik HL, Ferguson JD, Jones B. The effects of hypnotic susceptibility, suggestions for analgesia, and the utilization of cognitive strategies on the reduction of pain. *J Abnorm Psychol.* 1979;88:282–292.

20   Quartana PJ, Campbell CM, Edwards RR. Pain catastrophizing: a critical review. *Expert Rev Neurother.* 2009;9:745–758.

21   Darnall BD, Sturgeon JA, Cook KF, et al. Development and validation of a daily pain catastrophizing scale. *J Pain.* 2017;18:1139–1149.

22   Martire LM, Zhaoyang R, Marini CM, Nah S, Darnall BD. Daily and bidirectional linkages between pain catastrophizing and spouse responses. *Pain.* 2019;160:2841–2847; Stephens MAP, Martire LM, Cremeans-Smith JK, Druley JA, Wojno WC. Older women with osteoarthritis and their caregiving husbands: effects of pain and pain expression on husbands'

well-being and support. Washington, DC: American Psychological Association; 2006:3–12.

23 Martire LM, Keefe FJ, Schulz R, Parris Stephens MA, Mogle JA. The impact of daily arthritis pain on spouse sleep. *Pain*. 2013;154:1725–1731.

24 Klein LW, Tra Y, Garratt KN, et al. Occupational health hazards of interventional cardiologists in the current decade: results of the 2014 SCAI membership survey. *Catheter Cardio Inte*. 2015;86:913–924.

25 Grant M, O-Beirne-Elliman J, Froud R, Underwood M, Seers K. The work of return to work. Challenges of returning to work when you have chronic pain: a meta-ethnography. *BMJ Open*. 2019;9:e025743.

26 *Annual Statistical Report on the Social Security Disability Insurance Program, 2019* (Tables 39–45). 2020. https://www.ssa.gov/policy/docs/statcomps/di_asr/index.html.

27 *Annual Statistical Report on the Social Security Disability Insurance Program, 2019* (Tables 39–45). 2020. https://www.ssa.gov/policy/docs/statcomps/di_asr/index.html; Joffe-Walt C. Unfit for work: the startling rise of disability in America. NPR. 2013. https://www.npr.org/series/196621208/unfit-for-work-the-startling-rise-of-disability-in-america.

28 Brinjikji W, Luetmer PH, Comstock B, et al. Systematic literature review of imaging features of spinal degeneration in asymptomatic populations. *AJNR Am J Neuroradiol*. 2015;36:811–816.

29 Contorno S. Rand Paul says most people receive disability for back pain, anxiety. PolitiFact. January 16, 2015. https://www.politifact.com/factchecks/2015/jan/16/rand-paul/rand-paul-says-most-people-receive-disability-back.

CHAPTER 4. RAGE INSIDE THE MACHINE

1 5 things to know about Boston's hospital hotspot. *Stat News*. December 21, 2015. https://www.statnews.com/2015/12/21/longwood-medical-area. Accessed May 14, 2021.

2 Melzack R, Wall PD. Pain mechanisms: a new theory. *Science*. 1965;150:971–979.

3 Woolf CJ. Patrick D. Wall (1925–2001). *Nature*. 2001;413:378.

4 Squire LR, ed. *The History of Neuroscience in Autobiography*. San Diego, CA: Academic Press; 2001.

5 Squire LR, ed. *The History of Neuroscience in Autobiography*. San Diego, CA: Academic Press; 2001.

6   Melzack R, Wall PD. Pain mechanisms: a new theory. *Science*. 1965;150:971–979.

7   Katz J, Rosenbloom BN. The golden anniversary of Melzack and Wall's gate control theory of pain: celebrating 50 years of pain research and management. *Pain Res Manag*. 2015;20:285–286.

8   Squire LR, ed. *The History of Neuroscience in Autobiography*. San Diego, CA: Academic Press; 2001.

9   Murray JF, Schraufnagel DE, Hopewell PC. Treatment of tuberculosis: a historical perspective. *Ann Am Thorac Soc*. 2015;12:1749–1759.

10 Hicks CW, Selvin E. Epidemiology of peripheral neuropathy and lower extremity disease in diabetes. *Curr Diab Rep*. 2019;19:86.

11 Hippocrates of Cos. *The Sacred Disease*. Cambridge, MA: Harvard University Press; 1923.

12 Bourke J. Silas Weir Mitchell's The Case of George Dedlow. *Lancet*. 2009;373:1332–1333; Reilly RF. Medical and surgical care during the American Civil War, 1861–1865. *Proc (Bayl Univ Med Cent)*. 2016;29:138–142.

13 Army surgeons: their character and duties. In: Stevens EB, Murray JA, eds. *The Cincinnati Lancet and Observer*. Cincinnati, OH: E. B. Stevens, 1863:339.

14 Kline DG. Silas Weir Mitchell and "The Strange Case of George Dedlow." *Neurosurg Focus*. 2016;41:E5.

15 Puglionesi A. The Civil War doctor who proved phantom limb pain was real. History.com. Updated August 31, 2018. https://www.history .com/news/the-civil-war-doctor-who-proved-phantom-limb-pain-was -real. Accessed May 18, 2021.

16 Finger S, Hustwit MP. Five early accounts of phantom limb in context: Pare, Descartes, Lemos, Bell, and Mitchell. *Neurosurgery*. 2003;52:675–686; discussion 85–86.

17 Gabriel R. *Between Flesh and Steel: A History of Military Medicine from the Middle Ages to the War in Afghanistan*. Washington, DC: Potomac Books; 2013.

18 Finger S, Hustwit MP. Five early accounts of phantom limb in context: Pare, Descartes, Lemos, Bell, and Mitchell. *Neurosurgery*. 2003;52:675–686; discussion 85–86.

19 Descartes R. *The Philosophical Writings of Descartes*, Vol. 3: *The Correspondence*. Cottingham J, Stoothoff R, Murdoch D, Kenny A (trans). New York: Cambridge University Press; 1991.

20   Bailey AA, Moersch FP. Phantom limb. *Can Med Assoc J.* 1941;45:37–42.

21   Bailey AA, Moersch FP. Phantom limb. *Can Med Assoc J.* 1941;45:37–42.

22   Collins KL, Russell HG, Schumacher PJ, et al. A review of current theories and treatments for phantom limb pain. *J Clin Invest.* 2018;128: 2168–2176.

23   Collins KL, Russell HG, Schumacher PJ, et al. A review of current theories and treatments for phantom limb pain. *J Clin Invest.* 2018;128: 2168–2176.

24   Katz J, Melzack R. Pain "memories" in phantom limbs: review and clinical observations. *Pain.* 1990;43:319–336.

25   Nikolajsen L, Ilkjaer S, Kroner K, Christensen JH, Jensen TS. The influence of preamputation pain on postamputation stump and phantom pain. *Pain.* 1997;72:393–405; Melzack R, Israel R, Lacroix R, Schultz G. Phantom limbs in people with congenital limb deficiency or amputation in early childhood. *Brain.* 1997;120(Pt 9):1603–1620.

26   Saurat M, Agbakou M, Attigui P, Golmard J, Arnulf I. Walking dreams in congenital and acquired paraplegia. *Consciousness and Cognition.* 2011;20(4):1425–1432.

27   Bekrater-Bodmann R, Schredl M, Diers M, et al. Post-amputation pain is associated with the recall of an impaired body representation in dreams— results from a nation-wide survey on limb amputees. *PLoS One.* 2015;10:e0119552.

28   The curious case of the phantom penis. Vice.com. March 31, 2016. https://www.vice.com/en/article/vdxapx/the-curious-case-of-the-phantom -penis. Accessed May 20, 2021.

29   Woolf CJ. Evidence for a central component of post-injury pain hypersensitivity. *Nature.* 1983;306:686–688.

30   Hashmi JA, Baliki MN, Huang L, et al. Shape shifting pain: chronification of back pain shifts brain representation from nociceptive to emotional circuits. *Brain.* 2013;136:2751–2768; Tatu K, Costa T, Nani A, et al. How do morphological alterations caused by chronic pain distribute across the brain? A meta-analytic co-alteration study. *Neuroimage-Clin.* 2018;18: 15–30.

31   Ko HG, Kim JI, Sim SE, et al. The role of nuclear PKMzeta in memory maintenance. *Neurobiol Learn Mem.* 2016;135:50–56.

32   Li XY, Ko HG, Chen T, et al. Alleviating neuropathic pain hypersensitivity by inhibiting PKMzeta in the anterior cingulate cortex. *Science.* 2010;330:1400–1404.

33 Asiedu MN, Tillu DV, Melemedjian OK, et al. Spinal protein kinase M zeta underlies the maintenance mechanism of persistent nociceptive sensitization. *J Neurosci.* 2011;31:6646–6653.

34 BIM. Vania Apkarian and the holy grail. *Relief.* 2012. https://relief.news /2012/05/28/vania-apkarian-and-the-holy-grail. Accessed May 21, 2021.

35 Redelmeier DA, Kahneman D. Patients' memories of painful medical treatments: real-time and retrospective evaluations of two minimally invasive procedures. *Pain.* 1996;66:3–8.

36 Redelmeier DA, Katz J, Kahneman D. Memories of colonoscopy: a randomized trial. *Pain.* 2003;104:187–194.

37 Kensinger EA, Garoff-Eaton RJ, Schacter DL. How negative emotion enhances the visual specificity of a memory. *J Cogn Neurosci.* 2007;19:1872–1887; Noel M, Rabbitts JA, Tai GG, Palermo TM. Remembering pain after surgery: a longitudinal examination of the role of pain catastrophizing in children's and parents' recall. *Pain.* 2015;156:800–808; Noel M, Rabbitts JA, Fales J, Chorney J, Palermo TM. The influence of pain memories on children's and adolescents' post-surgical pain experience: a longitudinal dyadic analysis. *Health Psychol.* 2017;36:987–995.

38 Liu X, Liu Y, Li L, Hu Y, Wu S, Yao S. Overgeneral autobiographical memory in patients with chronic pain. *Pain Med.* 2014;15:432–439; Berger SE, Vachon-Presseau E, Abdullah TB, Baria AT, Schnitzer TJ, Apkarian AV. Hippocampal morphology mediates biased memories of chronic pain. *Neuroimage.* 2018;166:86–98.

39 Apkarian AV, Mutso AA, Centeno MV, et al. Role of adult hippocampal neurogenesis in persistent pain. *Pain.* 2016;157:418–428.

40 Dellarole A, Morton P, Brambilla R, et al. Neuropathic pain-induced depressive-like behavior and hippocampal neurogenesis and plasticity are dependent on TNFR1 signaling. *Brain Behav Immun.* 2014;41:65–81; Berger SE, Vachon-Presseau E, Abdullah TB, Baria AT, Schnitzer TJ, Apkarian AV. Hippocampal morphology mediates biased memories of chronic pain. *Neuroimage.* 2018;166:86–98; Oosterman JM, Hendriks H, Scott S, Lord K, White N, Sampson EL. When pain memories are lost: a pilot study of semantic knowledge of pain in dementia. *Pain Med.* 2014;15:751–757.

41 Chou R, Hartung D, Turner J, et al. Opioid treatments for chronic pain. Comparative Effectiveness Review No. 229. Agency for Healthcare Research and Quality, Rockville, MD. AHRQ Publication No 20-EHC011. 2020; Krebs EE, Gravely A, Nugent S, et al. Effect of opioid vs nonopioid

medications on pain-related function in patients with chronic back pain or hip or knee osteoarthritis pain: the SPACE Randomized Clinical Trial. *JAMA.* 2018;319(9):872–882; Wilson N, Kariisa M, Seth P, Smith H, IV, Davis NL. Drug and opioid-involved overdose deaths—United States, 2017–2018. *MMWR Morb Mortal Wkly Rep.* 2020;69:290–297.

CHAPTER 5. THE GOD OF DREAMS

1   Garrison G. Claire McCaskill cites disproven figure on opioid use. Politi-fact. May 10, 2017. https://www.politifact.com/factchecks/2017/may/10/claire-mccaskill/mccaskill-cites-long-disproven-figure-opioid-use. Accessed May 27, 2021; Boslett AJ, Denham A, Hill EL. Using contributing causes of death improves prediction of opioid involvement in unclassified drug over-doses in US death records. *Addiction.* 2020;115(7):1308–1317.

2   Institute of Medicine. *Relieving Pain in America: A Blueprint for Transform-ing Prevention, Care, Education, and Research.* Washington, DC: National Academies Press; 2011.

3   Inglis L. *Milk of Paradise: A History of Opium.* New York: Pegasus Books; 2019; Brook K, Bennett J, Desai SP. The chemical history of morphine: an 8000-year journey, from resin to de-novo synthesis. *J Anesth Hist.* 2017;3(2):50–55.

4   Kritikos PG, Papadaki, SP. The history of the poppy and of opium and their expansion in antiquity in the eastern Mediterranean area. *UNODC Bulle-tin on Narcotics.* 1967(3):17–38.

5   Kritikos PG, Papadaki, SP. The history of the poppy and of opium and their expansion in antiquity in the eastern Mediterranean area. *UNODC Bulle-tin on Narcotics.* 1967(3):17–38.

6   Kritikos PG, Papadaki, SP. The history of the poppy and of opium and their expansion in antiquity in the eastern Mediterranean area. *UNODC Bulle-tin on Narcotics.* 1967(3):17–38; Africa TW. The opium addiction of Marcus Aurelius. *J Hist Ideas.* 1961;22(1):97–102.

7   Inglis L. *Milk of Paradise: A History of Opium.* New York: Pegasus Books; 2019.

8   Littman G. "A splendid income": The world's greatest drug cartel. *Bilan.* November 24, 2015. https://www.bilan.ch/opinions/garry-littman/_a_splendid_income_the_world_s_greatest_drug_cartel. Accessed May 27, 2021.

9   Backhouse E, Bland, JOP. *Annals and Memoirs of the Court of Peking.* Boston: Houghton Mifflin; 1914.

10   Littman G. "A splendid income": The world's greatest drug cartel. *Bilan*. November 24, 2015. https://www.bilan.ch/opinions/garry-littman/_a_splendid _income_the_world_s_greatest_drug_cartel. Accessed May 27, 2021.

11   Lau J. Highlighting differences in interpretations of the Opium War. *New York Times*. August 18, 2011; Pardo B, Taylor J, Caulkins JP, Kilmer B, Reuter P, Stein BD. *The Future of Fentanyl and Other Synthetic Opioids*. Santa Monica, CA: RAND Corporation; 2019.

12   Trocki C. *Opium and Empire: Chinese Society in Colonial Singapore, 1800–1910*. Ithaca, NY: Cornell University Press; 1990.

13   Lyall A. The religious situation in India. In: *Asiatic Studies: Religious and Social*. London: John Murray, Albemarle Street; 1884.

14   Hacker JD. Recounting the dead. Opinionator. *New York Times*. September 20, 2011. https://opinionator.blogs.nytimes.com/2011/09/20/recounting -the-dead/#more-105317. Accessed May 27, 2021; Figg L, Farrell-Beck J. Amputation in the Civil War: physical and social dimensions. *J Hist Med Allied Sci*. 1993;48(4):454–475.

15   Brook K, Bennett J, Desai SP. The chemical history of morphine: an 8000-year journey, from resin to de-novo synthesis. *J Anesth Hist*. 2017;3(2):50–55.

16   Wood A. New method of treating neuralgia by the direct application of opiates to the painful points. *Edinb Med Surg J*. 1855;82(203): 265–281.

17   Howard-Jones N. A critical study of the origins and early development of hypodermic medication. *J Hist Med Allied Sci*. 1947;2(2):201–249.

18   Trickey E. Inside the story of America's 19th-century opiate addiction. *Smithsonian Magazine*. January 4, 2018. https://www.smithsonianmag.com /history/inside-story-americas-19th-century-opiate-addiction-180967673.

19   Daly JRL. A clinical study of heroin. *Boston Med Surg J*. 1900;142(8): 190–192.

20   Geiger HJ. Medical nemesis. *New York Times*. May 2, 1976.

21   Nunn R, Parsons J, Shambaugh J. A dozen facts about the economics of the U.S. health-care system. Brookings Institution. March 10, 2020. https://www.brookings.edu/research/a-dozen-facts-about-the-economics-of -the-u-s-health-care-system.

22   Mitchell EM. Concentration of health expenditures and selected characteristics of high spenders, U.S. civilian noninstitutionalized population, 2016. In: *Statistical Brief (Medical Expenditure Panel Survey [US])*. Rockville, MD: Agency for Healthcare Research and Quality (US); 2001;

Himmelstein DU, Jun M, Busse R, et al. A comparison of hospital administrative costs in eight nations: US costs exceed all others by far. *Health Aff (Millwood)* 2014;33(9):1586–1594.

23 *Pharmaceutical Drugs Global Market Report.* Bangalore: Business Research Company; 2018.

24 Macy B. *Dopesick: Dealers, Doctors, and the Drug Company That Addicted America.* Boston: Little, Brown; 2018.

CHAPTER 6. ANGEL OF MERCY

1 Fontana JS. The social and political forces affecting prescribing practices for chronic pain. *J Prof Nurs.* 2008;24(1):30–35.

2 Davies EC, Green CF, Taylor S, Williamson PR, Mottram DR, Pirmohamed M. Adverse drug reactions in hospital in-patients: a prospective analysis of 3695 patient-episodes. *PLoS One.* 2009;4(2):e4439.

3 Deng LX, Patel K, Miaskowski C, et al. Prevalence and characteristics of moderate to severe pain among hospitalized older adults. *J Am Geriatr Soc.* 2018;66(9):1744–1751; Desai R, Hong YR, Huo J. Utilization of pain medications and its effect on quality of life, health care utilization and associated costs in individuals with chronic back pain. *J Pain Res.* 2019;12:557–569.

4 Toye F, Seers K, Tierney S, Barker KL. A qualitative evidence synthesis to explore healthcare professionals' experience of prescribing opioids to adults with chronic non-malignant pain. *BMC Fam Pract.* 2017;18(1):94.

5 Li L, Setoguchi S, Cabral H, Jick S. Opioid use for noncancer pain and risk of myocardial infarction amongst adults. *J Intern Med.* 2013; 273(5):511–526; Fernando H, Nehme Z, Peter K, et al. Prehospital opioid dose and myocardial injury in patients with ST elevation myocardial infarction. *Open Heart.* 2020;7(2); Storey RF, Parker WAE. Opiates and clopidogrel efficacy: lost in transit? *J Am Coll Cardiol.* 2020;75(3):301–303; Bonin M, Mewton N, Roubille F, et al. Effect and safety of morphine use in acute anterior ST-segment elevation myocardial infarction. *J Am Heart Assoc.* 2018;7(4).

6 Institute of Medicine. *Relieving Pain in America: A Blueprint for Transforming Prevention, Care, Education, and Research.* Washington, DC: National Academies Press; 2011.

7 Fauber J. IOM and COI: painful disclosures? *Milwaukee Journal Sen-*

tinel/MedPage Today. June 25, 2014. https://www.medpagetoday.com /painmanagement/painmanagement/46482.

8   Humphreys K. Americans take more opioids than any other country— but not because they're in more pain. *Washington Post*. March 23, 2018. https://www.washingtonpost.com/news/wonk/wp/2018/03/23/americans -take-more-pain-pills-but-not-because-theyre-in-more-pain.

9   Trickey E. Inside the story of America's 19th-century opiate addiction. *Smithsonian Magazine*. January 4, 2018. https://www.smithsonianmag.com /history/inside-story-americas-19th-century-opiate-addiction-180967673.

10   6,000 opium users here. Dr. Hamilton Wright thinks five-sixths of them are white. *New York Times*. August 1, 1908; Marshall E. "Uncle Sam is the worst drug fiend in the world"; Dr Hamilton Wright, Opium Commissioner, says we use more of that drug per capita than the Chinese. *New York Times*. March 12, 1911.

11   Marshall E. "Uncle Sam is the worst drug fiend in the world"; Dr Hamilton Wright, Opium Commissioner, says we use more of that drug per capita than the Chinese. *New York Times*. March 12, 1911.

12   Marshall E. "Uncle Sam is the worst drug fiend in the world"; Dr Hamilton Wright, Opium Commissioner, says we use more of that drug per capita than the Chinese. *New York Times*. March 12, 1911.

13   Richmond C. Dame Cicely Saunders, founder of the modern hospice movement, dies. *Brit Med J*, 2005;331(7510):238; Stolberg S. A conversation with Dame Cicely Saunders; reflecting on a life of treating the dying. *New York Times*. May 11, 1999; Saunders C. *Watch with Me: Inspiration for a Life in Hospice Care*. Lancaster, UK: Observatory Publications; 2003.

14   Stolberg S. A conversation with Dame Cicely Saunders; reflecting on a life of treating the dying. *New York Times*. May 11, 1999.

15   Saunders C. *Watch with Me: Inspiration for a Life in Hospice Care*. Lancaster, UK: Observatory Publications; 2003.

16   Saunders C. Dying of cancer. *St Thomas's Hospital Gazette*. 1958;56(2): 37–47.

17   Stolberg S. A conversation with Dame Cicely Saunders; reflecting on a life of treating the dying. *New York Times*. May 11, 1999.

18   Saunders C. A personal therapeutic journey. *BMJ*. 1996;313(7072):1599– 1601.

19   Saunders C. The treatment of intractable pain in terminal cancer. *Proc R Soc Med*. 1963;56:195–197.

20   Jeffrey D. *Against Physician Assisted Suicide: A Palliative Care Perspective*. New York: Radcliffe Publishing; 2009.

21   Saunders C. The treatment of intractable pain in terminal cancer. *Proc R Soc Med*. 1963;56:195–197.

22   Saunders C. The treatment of intractable pain in terminal cancer. *Proc R Soc Med*. 1963;56:195–197.

23   Stolberg S. A conversation with Dame Cicely Saunders; reflecting on a life of treating the dying. *New York Times*. May 11, 1999.

24   Saunders C. The treatment of intractable pain in terminal cancer. *Proc R Soc Med*. 1963;56:195–197.

25   Posner G. *Pharma: Greed, Lies, and the Poisoning of America*. New York: Simon & Schuster; 2021; Reynolds LA, Tansey EM, eds. *Innovation in Pain Management*. London: Wellcome Trust Centre for the History of Medicine at UCL; 2004.

26   Leech AA, Cooper WO, McNeer E, Scott TA, Patrick SW. Neonatal abstinence syndrome in the United States, 2004–16. *Health Aff (Millwood)*. 2020;39(5):764–767; Villapiano NL, Winkelman TN, Kozhimannil KB, Davis MM, Patrick SW. Rural and urban differences in neonatal abstinence syndrome and maternal opioid use, 2004 to 2013. *JAMA Pediatr*. 2017;171(2):194–196.

27   Keefe PR. *Empire of Pain: The Secret History of the Sackler Dynasty*. New York: Doubleday; 2021.

28   Young JH. *The Toadstool Millionaires: A Social History of Patent Medicines in America Before Federal Regulation*. Princeton, NJ: Princeton University Press; 1961.

29   Younkin P. Making the market: how the American pharmaceutical industry transformed itself during the 1940s. UC Berkeley: Center for Culture, Organizations and Politics—Previously Affiliated. 2008. https://escholarship.org/uc/item/2g67r185.

30   Fugh-Berman A, Alladin K, Chow J. Advertising in medical journals: Should current practices change? *PLoS Med*. 2006;3(6):e130; Harris R. *The Real Voice*. New York: Macmillan; 1964.

31   Lewin T. Drug makers fighting back against generics. *New York Times*. July 28, 1987, https://www.nytimes.com/1987/07/28/business/drug-makers-fighting-back-against-advance-of-generics.html; Donohue J. A history of drug advertising: the evolving roles of consumers and consumer protection. *Milbank Q*. 2006;84(4):659–699.

32   Keefe PR. The family that built an empire of pain. *New Yorker*. October

30, 2017. https://www.newyorker.com/magazine/2017/10/30/the-family-that
-built-an-empire-of-pain.

33   Ransom A. Direct-to-physician advertising, depression, and the advertised female patient. PhD diss., University of Tennessee; 2017; Podolsky SH, Herzberg D, Greene JA. Preying on prescribers (and their patients)— pharmaceutical marketing, iatrogenic epidemics, and the Sackler legacy. *N Engl J Med.* 2019;380(19):1785–1787.

34   Bobrow RS. Selections from current literature: benzodiazepines revisited. *Fam Pract.* 2003;20(3):347–349.

35   Dougherty P. Advertising; generic drugs and agencies. *New York Times.* September 12, 1985. https://www.nytimes.com/1985/09/12/business /advertising-generic-drugs-and-agencies.html.

36   Applbaum K. Pharmaceutical marketing and the invention of the medical consumer. *PLoS Med.* 2006;3(4):e189.

37   Tuttle AH, Tohyama S, Ramsay T, et al. Increasing placebo responses over time in U.S. clinical trials of neuropathic pain. *Pain.* 2015;156(12):2616–2626.

38   Keefe PR. The family that built an empire of pain. *New Yorker.* October 30, 2017. https://www.newyorker.com/magazine/2017/10/30/the-family -that-built-an-empire-of-pain.

39   Seth P, Scholl L, Rudd RA, Bacon S. Overdose deaths involving opioids, cocaine, and psychostimulants—United States, 2015–2016. *MMWR Morb Mortal Wkly Rep.* 2018;67(12):349–358.

40   *2018 Annual Surveillance Report of Drug-Related Risks and Outcomes— United States. Surveillance Special Report.* Atlanta: Centers for Disease Control and Prevention, US Department of Health and Human Services; 2018.

41   Chou R, Hartung D, Turner J, et al. Opioid treatments for chronic pain. Comparative Effectiveness Review No. 229. Agency for Healthcare Research and Quality, Rockville, MD. AHRQ Publication No 20-EHC011. 2020; Krebs EE, Gravely A, Nugent S, et al. Effect of opioid vs nonopioid medications on pain-related function in patients with chronic back pain or hip or knee osteoarthritis pain: the SPACE Randomized Clinical Trial. *JAMA.* 2018;319(9):872–882.

42   Grace PM, Strand KA, Galer EL, et al. Morphine paradoxically prolongs neuropathic pain in rats by amplifying spinal NLRP3 inflammasome activation. *Proc Natl Acad Sci USA.* 2016;113(24):E3441–E3450; Carroll CP, Lanzkron S, Haywood C, Jr., et al. Chronic opioid therapy and

central sensitization in sickle cell disease. *Am J Prev Med*. 2016;51(1 Suppl 1):S69–S77; Rivat C, Ballantyne J. The dark side of opioids in pain management: basic science explains clinical observation. *Pain Rep*. 2016;1(2):e570; Yi P, Pryzbylkowski P. Opioid induced hyperalgesia. *Pain Med*. 2015;16(Suppl 1):S32–S36.

CHAPTER 7. CROWN OF THORNS

1   Coffin PO, Rowe C, Oman N, et al. Illicit opioid use following changes in opioids prescribed for chronic non-cancer pain. *PLoS One*. 2020; 15(5):e0232538.

2   Monico LB, Mitchell SG. Patient perspectives of transitioning from prescription opioids to heroin and the role of route of administration. *Subst Abuse Treat Prev Policy*. 2018;13(1):4; Carlson RG, Nahhas RW, Martins SS, Daniulaityte R. Predictors of transition to heroin use among initially non-opioid dependent illicit pharmaceutical opioid users: a natural history study. *Drug Alcohol Depend*. 2016;160:127–134.

3   Kadri AN, Wilner B, Hernandez AV, et al. Geographic trends, patient characteristics, and outcomes of infective endocarditis associated with drug abuse in the United States from 2002 to 2016. *J Am Heart Assoc*. 2019;8(19):e012969.

4   Dreborg S, Sundstrom G, Larsson TA, Larhammar D. Evolution of vertebrate opioid receptors. *Proc Natl Acad Sci USA*. 2008;105(40):15487–15492; Stein C, Hassan AH, Lehrberger K, Giefing J, Yassouridis A. Local analgesic effect of endogenous opioid peptides. *Lancet*. 1993;342(8867): 321–324.

5   Janssen SA, Arntz A. Real-life stress and opioid-mediated analgesia in novice parachute jumpers. *J Psychophysiol*. 2001;15(2):106–113; Valentino RJ, Van Bockstaele E. Endogenous opioids: the downside of opposing stress. *Neurobiol Stress*. 2015;1:23–32.

6   Valentino RJ, Van Bockstaele E. Endogenous opioids: opposing stress with a cost. *F1000Prime Rep*. 2015;7:58. doi:10.12703/P7-58.

7   Chou R, Hartung D, Turner J, et al. Opioid treatments for chronic pain. Comparative Effectiveness Review No. 229. Agency for Healthcare Research and Quality, Rockville, MD. AHRQ Publication No 20-EHC011. 2020.

8   Fricker LD, Margolis EB, Gomes I, Devi LA. Five decades of research on opioid peptides: current knowledge and unanswered questions. *Mol Pharmacol*. 2020;98(2):96–108.

9   Martel MO, Petersen K, Cornelius M, Arendt-Nielsen L, Edwards R. Endogenous pain modulation profiles among individuals with chronic pain: relation to opioid use. *J Pain.* 2019;20(4):462–471.

10 Xu GP, Van Bockstaele E, Reyes B, Bethea T, Valentino RJ. Chronic morphine sensitizes the brain norepinephrine system to corticotropin-releasing factor and stress. *J Neurosci.* 2004;24(38):8193–8197; Sullivan MD. Depression effects on long-term prescription opioid use, abuse, and addiction. *Clin J Pain.* 2018;34(9):878–884.

11 Inagaki TK, Hazlett LI, Andreescu C. Opioids and social bonding: effect of naltrexone on feelings of social connection and ventral striatum activity to close others. *J Exp Psychol Gen.* 2020;149(4):732–745.

12 Paredes RG. Opioids and sexual reward. *Pharmacol Biochem Behav.* 2014;121:124–131; Koob GF. Neurobiology of opioid addiction: opponent process, hyperkatifeia, and negative reinforcement. *Biol Psychiatry.* 2020;87(1):44–53.

13 Lopez-Martinez AE, Reyes-Perez A, Serrano-Ibanez ER, Esteve R, Ramirez-Maestre C. Chronic pain, posttraumatic stress disorder, and opioid intake: a systematic review. *World J Clin Cases.* 2019;7(24):4254–4269.

14 Porter J, Jick H. Addiction rare in patients treated with narcotics. *N Engl J Med.* 1980;302(2):123.

15 Haney T, Hsu, A. Doctor who wrote 1980 letter on painkillers regrets that it fed the opioid crisis. NPR. June 16, 2017. https://www.npr.org/sections/health-shots/2017/06/16/533060031/doctor-who-wrote-1980-letter-on-pain killers-regrets-that-it-fed-the-opioid-crisi. Accessed June 2, 2021.

16 Leung PTM, Macdonald EM, Stanbrook MB, Dhalla IA, Juurlink DN. A 1980 letter on the risk of opioid addiction. *N Engl J Med.* 2017;376(22):2194–2195.

17 Portenoy RK, Foley KM. Chronic use of opioid analgesics in nonmalignant pain: report of 38 cases. *Pain.* 1986;25(2):171–186.

18 Lurie J. Unsealed documents show how Purdue Pharma created a "pain movement." *Mother Jones.* August 29, 2019. https://www.motherjones.com/crime-justice/2019/08/unsealed-documents-show-how-purdue-pharma-created-a-pain-movement. Accessed June 2, 2021.

19 *Fueling an Epidemic: Exposing the Financial Ties Between Opioid Manufacturers and Third Party Advocacy Groups (Report 2).* US Senate Homeland Security and Governmental Affairs Committee; 2018.

20 Perrone M. Federal pain panel rife with links to pharma companies.

Associated Press. January 27, 2016. https://apnews.com/article/6e22f8 ffcded4b2e9ba278bcc00f4f53.

21 Offices of Representatives Katherine Clark and Hal Rogers. *Corrupting In-fluence: Purdue and the WHO. Report: Exposing Dangerous Opioid Manufacturer Influence at the World Health Organization.* Bethesda, MD: ProQuest; 2019.

22 Amended Complaint, *State of Connecticut v. Purdue Pharma LP et al.*, No. X07 HHD-CV-19-6105325-S. Connecticut Superior Court; 2019.

23 Brower V. Patients see proposed FDA opioid rules as painfully restrictive. *Nat Med.* 2009;15(8):827; Opioid makers not fully responsible for abuse prevention, says FDA panel. *The Pharma Letter.* November 2, 2002. https://www.thepharmaletter.com/article/opioid-makers-not-fully-responsible-for-abuse-prevention-says-fda-panel. Accessed June 2, 2021.

24 Lurie J. Unsealed documents show how Purdue Pharma created a "pain movement." *Mother Jones.* August 29, 2019. https://www.motherjones.com/crime-justice/2019/08/unsealed-documents-show-how-purdue-pharma-created-a-pain-movement. Accessed June 2, 2021.

25 Weissman DE, Haddox DJ. Opioid pseudoaddiction—an iatrogenic syndrome. *Pain.* 1989;36(3):363–366; Greene MS, Chambers, RA. Pseudoaddiction: fact or fiction? An investigation of the medical literature. *Curr Addict Rep.* 2015;2(4):310–317; Passik SD, Kirsh KL, Webster L. Pseudoaddiction revisited: a commentary on clinical and historical considerations. *Pain Manag.* 2011;1(3):239–248.

26 Esch C. How one sentence helped set off the opioid crisis. *Marketplace.* December 13, 2017. https://www.marketplace.org/2017/12/13/opioid. Accessed June 3, 2021.

27 The marketing of OxyContin®: a cautionary tale. *Indian J Med Ethics.* 2019;4(3):183–193.

28 The marketing of OxyContin®: a cautionary tale. *Indian J Med Ethics.* 2019;4(3):183–193; Ryan H, Girion, L, Glover, S. "You want a description of hell?" OxyContin's 12-hour problem. *LA Times.* May 5, 2016. https://www.latimes.com/projects/oxycontin-part1; Whitaker B. Did the FDA ignite the opioid epidemic? *60 Minutes.* February 24, 2019. https://www.cbsnews.com/news/opioid-epidemic-did-the-fda-ignite-the-crisis-60-minutes. Accessed June 3, 2021.

29 Sullivan L. Critics question FDA's approval of Zohydro. NPR. February 26, 2014. https://www.npr.org/2014/02/26/282836473/critics-question-fdas-approval-of-zohydro. Accessed June 3, 2021.

30　Ahmad FB, Anderson RN. The leading causes of death in the US for 2020. *JAMA*. 2021;325(18):1829–1830.

31　Groenewald CB, Murray CB, Palermo TM. Adverse childhood experiences and chronic pain among children and adolescents in the United States. *Pain Rep*. 2020;5(5):e839.

CHAPTER 8. THE PAIN OF THE POWERLESS

1　Goyal MK, Kuppermann N, Cleary SD, Teach SJ, Chamberlain JM. Racial disparities in pain management of children with appendicitis in emergency departments. *JAMA Pediatr*. 2015;169(11):996–1002.

2　Flogging. Jewish Virtual Library. https://www.jewishvirtuallibrary.org /flogging. Accessed June 7, 2021; "Religion." *Encyclopedia Britannica*. February 2, 2021. https://www.britannica.com/topic/religion. Accessed June 7, 2021.

3　Schrader AM. Containing the spectacle of punishment: the Russian autocracy and the abolition of the knout, 1817–1845. *Slavic Review*. 1997;56(4):613–644; Langley H. *Social Reform in the U.S. Navy, 1798–1862*. Urbana, IL: University of Illinois Press; 1967.

4　Evans M, Morgan, R. *Preventing Torture: A Study of the European Convention for the Prevention of Torture and Inhuman or Degrading Treatment or Punishment*. Oxford, UK: Clarendon Press; 1998; Demosthenes. *Orations, Vol. IV: Orations 27–40: Private Cases*. Murray AT, trans. London: W. Heinemann Ltd.; 1936.

5　Maio G. History of medical involvement in torture—then and now. *Lancet*. 2001;357(9268):1609–1611; duBois P. *Torture and Truth*. London: Routledge Revivals; 1991.

6　Einolf CJ. The fall and rise of torture: a comparative and historical analysis. *Sociological Theory*. 2007;25(2):101–121.

7　Silverman L. *Tortured Subjects: Pain, Truth, and the Body in Early Modern France*. Chicago: University of Chicago Press; 2001.

8　Bourke J. Pain sensitivity: an unnatural history from 1800 to 1965. *J Med Humanit*. 2014;35(3):301–319; Royster HA. A review of the operations at St. Agnes Hospital, with remarks upon surgery in the Negro. *J Natl Med Assoc*. 1914;6(4):221–225.

9　Galton F. *Inquiries into Human Faculty and Its Development*. London: Macmillan; 1883; Mitchell SW. Civilization and pain. *J Am Med Assoc*. 1892;18(108).

10 Bourke J. Pain sensitivity: an unnatural history from 1800 to 1965. *J Med Humanit*. 2014;35(3):301–319.

11 Toledo AH. The medical legacy of Benjamin Rush. *J Invest Surg*. 2004;17(2):61–63.

12 Myers B. *"Drapetomania": Rebellion, Defiance and Free Black Insanity in the Antebellum United States*. Los Angeles: University of California, Los Angeles; 2014; Cartwright SA. How to save the republic, and the position of the South in the Union. *De Bow's Review, Agricultural, Commercial, Industrial Progress and Resources*. 1851;11(2).

13 Frank J. Sympathy and separation: Benjamin Rush and the contagious public. *Modern Intellectual History*. 2009;6(1):27–57; Cartwright SA. Report on the diseases and physical peculiarities of the Negro race. *New Orleans Med Surg J*. 1851;7:691–715.

14 Cartwright SA. Report on diseases and peculiarities of the Negro race. *De Bow's Review, Agricultural, Commercial, Industrial Progress and Resources*. 1851;11(3).

15 Cartwright SA. Unity of the human race disproved by the Hebrew Bible. *De Bow's Review, Agricultural, Commercial, Industrial Progress and Resources*. 1860;29(2):129–136.

16 Hoffman KM, Trawalter S, Axt JR, Oliver MN. Racial bias in pain assessment and treatment recommendations, and false beliefs about biological differences between blacks and whites. *Proc Natl Acad Sci USA*. 2016;113(16):4296–4301.

17 Chibnall JT, Tait RC. Disparities in occupational low back injuries: predicting pain-related disability from satisfaction with case management in African Americans and Caucasians. *Pain Med*. 2005;6(1):39–48; Chibnall JT, Tait RC, Andresen EM, Hadler NM. Race differences in diagnosis and surgery for occupational low back injuries. *Spine (Phila Pa 1976)*. 2006;31(11):1272–1275; Anderson KO, Green CR, Payne R. Racial and ethnic disparities in pain: causes and consequences of unequal care. *J Pain*. 2009;10(12):1187–1204.

18 Bijur P, Berard A, Esses D, Calderon Y, Gallagher EJ. Race, ethnicity, and management of pain from long-bone fractures: a prospective study of two academic urban emergency departments. *Acad Emerg Med*. 2008;15(7):589–597; Tamayo-Sarver JH, Hinze SW, Cydulka RK, Baker DW. Racial and ethnic disparities in emergency department analgesic prescription. *Am J Public Health*. 2003;93(12):2067–2073; Mills AM, Shofer FS, Boulis AK, Holena DN, Abbuhl SB. Racial disparity in analgesic treat-

ment for ED patients with abdominal or back pain. *Am J Emerg Med.* 2011;29(7):752–756; Alexander MJ, Kiang MV, Barbieri M. Trends in Black and white opioid mortality in the United States, 1979–2015. *Epidemiology.* 2018;29(5):707–715.

19   Netherland J, Hansen H. White opioids: pharmaceutical race and the war on drugs that wasn't. *Biosocieties.* 2017;12(2):217–238.

20   Ung E. In neighborhoods, mourning the lives lost to a legal drug. *Philadelphia Inquirer.* July 31, 2001.

21   Netherland J, Hansen H. White opioids: pharmaceutical race and the war on drugs that wasn't. *Biosocieties.* 2017;12(2):217–238.

22   Netherland J, Hansen H. White opioids: pharmaceutical race and the war on drugs that wasn't. *Biosocieties.* 2017;12(2):217–238; Larochelle MR, Bernson D, Land T, et al. Medication for opioid use disorder after nonfatal opioid overdose and association with mortality: a cohort study. *Ann Intern Med.* 2018;169(3):137–145; Goedel WC, Shapiro A, Cerda M, Tsai JW, Hadland SE, Marshall BDL. Association of racial/ethnic segregation with treatment capacity for opioid use disorder in counties in the United States. *JAMA Netw Open.* 2020;3(4):e203711; Lagisetty PA, Ross R, Bohnert A, Clay M, Maust DT. Buprenorphine treatment divide by race/ethnicity and payment. *JAMA Psychiat.* 2019;76(9):979–981.

23   Kim HJ, Yang GS, Greenspan JD, et al. Racial and ethnic differences in experimental pain sensitivity: systematic review and meta-analysis. *Pain.* 2017;158(2):194–211; Campbell CM, Edwards RR, Fillingim RB. Ethnic differences in responses to multiple experimental pain stimuli. *Pain.* 2005;113(1–2):20–26.

24   Begley S. Brain-imaging study shows "hot spot" that may explain why African Americans feel greater pain. *Stat News.* February 3, 2020. https://www .statnews.com/2020/02/03/brain-imaging-study-may-explain-why-african -americans-feel-greater-pain. Accessed June 9, 2021; Wager TD, Atlas LY, Lindquist MA, Roy M, Woo CW, Kross E. An fMRI-based neurologic signature of physical pain. *N Engl J Med.* 2013;368(15):1388–1397.

25   Losin EAR, Woo CW, Medina NA, Andrews-Hanna JR, Eisenbarth H, Wager TD. Neural and sociocultural mediators of ethnic differences in pain. *Nat Hum Behav.* 2020;4(5):517–530.

26   Hutcherson CA, Plassmann H, Gross JJ, Rangel A. Cognitive regulation during decision making shifts behavioral control between ventromedial and dorsolateral prefrontal value systems. *J Neurosci.* 2012;32(39):13543–

13554; Sokol-Hessner P, Camerer CF, Phelps EA. Emotion regulation reduces loss aversion and decreases amygdala responses to losses. *Soc Cogn Affect Neurosci.* 2013;8(3):341–350; Baliki MN, Geha PY, Fields HL, Apkarian AV. Predicting value of pain and analgesia: nucleus accumbens response to noxious stimuli changes in the presence of chronic pain. *Neuron.* 2010;66(1):149–160.

27   Kim HJ, Yang GS, Greenspan JD, et al. Racial and ethnic differences in experimental pain sensitivity: systematic review and meta-analysis. *Pain.* 2017;158(2):194–211; Campbell CM, Edwards RR, Fillingim RB. Ethnic differences in responses to multiple experimental pain stimuli. *Pain.* 2005;113(1–2):20–26.

28   Anderson SR, Gianola M, Perry JM, Losin EAR. Clinician-patient racial/ethnic concordance influences racial/ethnic minority pain: evidence from simulated clinical interactions. *Pain Med.* 2020;21(11):3109–3125.

29   Elliott ML, Knodt AR, Ireland D, et al. What is the test-retest reliability of common task-functional MRI measures? New empirical evidence and a meta-analysis. *Psychol Sci.* 2020;31(7):792–806; Goodin BR, Pham QT, Glover TL, et al. Perceived racial discrimination, but not mistrust of medical researchers, predicts the heat pain tolerance of African Americans with symptomatic knee osteoarthritis. *Health Psychol.* 2013;32(11):1117–1126; Ziadni MS, Sturgeon JA, Bissell D, et al. Injustice appraisal, but not pain catastrophizing, mediates the relationship between perceived ethnic discrimination and depression and disability in low back pain. *J Pain.* 2020;21(5–6):582–592.

30   Diversity in medicine: facts and figures 2019. Figure 18. Percentage of all active physicians by race/ethnicity, 2018. AAMC. https://www.aamc.org/data-reports/workforce/interactive-data/figure-18-percentage-all-active-physicians-race/ethnicity-2018. Accessed June 9, 2021.

31   Burger O, Baudisch A, Vaupel JW. Human mortality improvement in evolutionary context. *Proc Natl Acad Sci USA.* 2012;109(44):18210–18214; Warraich HJ, Califf RM. Differences in health outcomes between men and women: biological, behavioral, and societal factors. *Clin Chem.* 2019;65(1):19–23.

32   Steingrimsdottir OA, Landmark T, Macfarlane GJ, Nielsen CS. Defining chronic pain in epidemiological studies: a systematic review and meta-analysis. *Pain.* 2017;158(11):2092–2107; Barker KK. Listening to Lyrica: contested illnesses and pharmaceutical determinism. *Soc Sci Med.* 2011;73(6):833–842.

33   Mogil JS. Qualitative sex differences in pain processing: emerging evidence of a biased literature. *Nat Rev Neurosci.* 2020;21(7):353–365; Vambheim SM, Flaten MA. A systematic review of sex differences in the placebo and the nocebo effect. *J Pain Res.* 2017;10:1831–1839; Hoffmann DE, Tarzian AJ. The girl who cried pain: a bias against women in the treatment of pain. *J Law Med Ethics.* 2001;29(1):13–27; McDonald DD, Bridge RG. Gender stereotyping and nursing care. *Res Nurs Health.* 1991;14(5):373–378; Steingrimsdottir OA, Landmark T, Macfarlane GJ, Nielsen CS. Defining chronic pain in epidemiological studies: a systematic review and meta-analysis. *Pain.* 2017;158(11):2092–2107.

34   Shipman PL. Why is human childbirth so painful? *Am Sci.* 2013;101(6); Albers LL. The duration of labor in healthy women. *J Perinatol.* 1999;19(2):114–119.

35   Sadatmoosavi Z, Shokouhi MA. Relationship of Eve's sin myth in Judeo-Christian and Islamic teachings with women education. 2nd International Seminar of Teaching Excellence and Innovation. February 25, 2014.

36   Helmuth L. The disturbing, shameful history of childbirth deaths. *Slate.* September 10, 2013. https://slate.com/technology/2013/09/death-in-childbirth-doctors-increased-maternal-mortality-in-the-20th-century-are-midwives-better.html. Accessed June 9, 2021.

37   Lurie S. Euphemia Maclean, Agnes Sampson and pain relief during labour in 16th century Edinburgh. *Anaesthesia.* 2004;59(8):834–835.

38   Hibbard B. *The Obstetrician's Armamentarium: Historical Obstetric Instruments and Their Inventors.* Novato, CA: Norman Publishing; 2000.

39   Dunn PM. Sir James Young Simpson (1811–1870) and obstetric anaesthesia. *Arch Dis Child Fetal Neonatal Ed.* 2002;86(3):F207–F209.

40   Caton D. *What a Blessing She Had Chloroform: The Medical and Social Response to the Pain of Childbirth from 1800 to the Present.* New Haven, CT: Yale University Press; 1999.

41   Simpson JA. On a new anaesthetic agent, more efficient than sulphuric ether. *Lancet.* 1847;50(1264):549–550.

42   Caton D. *What a Blessing She Had Chloroform: The Medical and Social Response to the Pain of Childbirth from 1800 to the Present.* New Haven, CT: Yale University Press; 1999; Barry E. Chloroform in childbirth? Yes, please, the queen said. *New York Times.* May 6, 2019. https://www.nytimes.com/2019/05/06/world/europe/uk-royal-births-labor.html.

43   Barry E. Chloroform in childbirth? Yes, please, the queen said. *New York*

Times. May 6, 2019. https://www.nytimes.com/2019/05/06/world/europe
/uk-royal-births-labor.html.

44 Finkbeiner A. Labor dispute. *New York Times*. October 31, 1999.
    https://www.nytimes.com/1999/10/31/books/labor-dispute.html.

45 Skowronski GA. Pain relief in childbirth: changing historical and feminist
    perspectives. *Anaesth Intens Care*. 2015;43Suppl:25–28; MacIvor Thomp-
    son L. The politics of female pain: women's citizenship, twilight sleep and
    the early birth control movement. *Med Humanit*. 2019;45(1):67–74.

46 MacIvor Thompson L. The politics of female pain: women's citizenship,
    twilight sleep and the early birth control movement. *Med Humanit*.
    2019;45(1):67–74.

47 Bryan WT. On American motherhood. Theodore Roosevelt. In: Halsey F,
    ed. *The World's Famous Orations*. New York: Funk and Wagnalls; 1906.

48 Leavitt JW. Birthing and anesthesia: the debate over twilight sleep. *Signs*.
    1980;6(1):147–164.

49 Loeser JD. In memoriam: John J. Bonica (1917–1994). *Pain*. 1994;59(7).

50 Collins M. Natural birth. Lucie's List. https://www.lucieslist.com/natural
    -birth/#gref. Updated April 2021. Accessed June 9, 2021.

51 Dick-Read G. *Motherhood in the Post-war World*. London: Heinemann
    Medical Books; 1944; Dick-Read G. *Revelation of Childbirth: The Princi-
    ples and Practice of Natural Childbirth*. London: Heinemann Medical
    Books; 1942.

52 Slee E. Misogyny and the epidural: a primer. Medium. November 26, 2018.
    https://medium.com/s/story/misogyny-and-the-epidural-a-primer-7749328e8
    999. Accessed June 9, 2021.

53 Badreldin N, Grobman WA, Yee LM. Racial disparities in postpartum pain
    management. *Obstet Gynecol*. 2019;134(6):1147–1153; Leal MDC, Gama
    S, Pereira APE, Pacheco VE, Carmo CND, Santos RV. The color of pain:
    racial iniquities in prenatal care and childbirth in Brazil. *Cad Saude Pub-
    lica*. 2017;33(Suppl 1):e00078816; Souza MA, Guida JPS, Cecatti JG, et
    al. Analgesia during labor and vaginal birth among women with severe
    maternal morbidity: secondary analysis from the WHO Multicountry Sur-
    vey on Maternal and Newborn Health. *Biomed Res Int*. 2019;2019:
    7596165.

54 Zelaya CE, Dahlhamer JM, Lucas JW, Connor EM. Chronic pain and
    high-impact chronic pain among U.S. adults, 2019. *NCHS Data Brief*.
    2020(390):1–8; Mogil JS. Qualitative sex differences in pain processing:
    emerging evidence of a biased literature. *Nat Rev Neurosci*.

2020;21(7):353–365; Hoffmann DE, Tarzian AJ. The girl who cried pain: a bias against women in the treatment of pain. *J Law Med Ethics.* 2001;29(1):13–27; Sorenson SB. Gender disparities in injury mortality: consistent, persistent, and larger than you'd think. *Am J Public Health.* 2011;101(Suppl 1):S353–S358.

55   Mogil JS, Chanda ML. The case for the inclusion of female subjects in basic science studies of pain. *Pain.* 2005;117(1–2):1–5.

56   Steiner TJ, Stovner LJ, Jensen R, Uluduz D, Katsarava Z. Migraine remains second among the world's causes of disability, and first among young women: findings from GBD2019. *J Headache Pain.* 2020;21(1):137.

57   Borsook D, Erpelding N, Lebel A, et al. Sex and the migraine brain. *Neurobiol Dis.* 2014;68:200–214; Hau M, Dominguez OA, Evrard HC. Testosterone reduces responsiveness to nociceptive stimuli in a wild bird. *Horm Behav.* 2004;46(2):165–170; Basaria S, Travison TG, Alford D, et al. Effects of testosterone replacement in men with opioid-induced androgen deficiency: a randomized controlled trial. *Pain.* 2015;156(2):280–288; Aloisi AM, Bachiocco V, Costantino A, et al. Cross-sex hormone administration changes pain in transsexual women and men. *Pain.* 2007;132Suppl 1:S60–S67. doi: 10.1016/j.pain.2007.02.006.

58   Avona A, Burgos-Vega C, Burton MD, Akopian AN, Price TJ, Dussor G. Dural calcitonin gene-related peptide produces female-specific responses in rodent migraine models. *J Neurosci.* 2019;39(22):4323–4331; Cetinkaya A, Kilinc E, Camsari C, Ogun MN. Effects of estrogen and progesterone on the neurogenic inflammatory neuropeptides: implications for gender differences in migraine. *Exp Brain Res.* 2020;238(11):2625–2639.

59   Cohen J. Migraine breakthrough: not so fast. Forbes.com. June 6, 2018. https://www.forbes.com/sites/joshuacohen/2018/06/06/migraine-breakthrough-not-so-fast/?sh=35ab46fe8971. Accessed June 9, 2021.

60   Samulowitz A, Gremyr I, Eriksson E, Hensing G. "Brave men" and "emotional women": a theory-guided literature review on gender bias in health care and gendered norms towards patients with chronic pain. *Pain Res Manag.* 2018;2018:6358624.

61   Samulowitz A, Gremyr I, Eriksson E, Hensing G. "Brave men" and "emotional women": a theory-guided literature review on gender bias in health care and gendered norms towards patients with chronic pain. *Pain Res Manag.* 2018;2018:6358624; Vincent A, Lahr BD, Wolfe F, et al. Prevalence of fibromyalgia: a population-based study in Olmsted County,

Minnesota, utilizing the Rochester Epidemiology Project. *Arthritis Care Res.* 2013;65:786–792. https://doi.org/10.1002/acr.21896.

62   Munce SE, Stewart DE. Gender differences in depression and chronic pain conditions in a national epidemiologic survey. *Psychosomatics.* 2007;48(5):394–399.

63   Serdarevic M, Striley CW, Cottler LB. Sex differences in prescription opioid use. *Curr Opin Psychiatry.* 2017;30(4):238–246; Cicero TJ, Wong G, Tian Y, Lynskey M, Todorov A, Isenberg K. Co-morbidity and utilization of medical services by pain patients receiving opioid medications: data from an insurance claims database. *Pain.* 2009;144(1–2):20–27; Schieber LZ, Guy GP, Jr., Seth P, Losby JL. Variation in adult outpatient opioid prescription dispensing by age and sex—United States, 2008–2018. *MMWR Morb Mortal Wkly Rep.* 2020;69(11):298–302.

64   Doshi TL, Bicket MC. Why aren't there more female pain medicine physicians? *Reg Anesth Pain Med.* 2018;43(5):516–520; AAPM Board of Directors. American Association of Pain Medicine. https://painmed.org /board-of-directors. Accessed June 11, 2021.

65   Doshi TL, Bicket MC. Why aren't there more female pain medicine physicians? *Reg Anesth Pain Med.* 2018;43(5):516–520.

66   Doctor gets $7m in gender bias suit. Boston.com. www.boston.com /culture/health/2013/02/06/doctor-gets-7m-in-gender-bias-suit/. Published 2013. Accessed June 11, 2021.

67   Jefferson L, Bloor K, Birks Y, Hewitt C, Bland M. Effect of physicians' gender on communication and consultation length: a systematic review and meta-analysis. *J Health Serv Res Policy.* 2013;18(4):242–248; Bertakis KD, Azari R. Patient-centered care: the influence of patient and resident physician gender and gender concordance in primary care. *J Womens Health (Larchmt).* 2012;21(3):326–333; Hirsh AT, Hollingshead NA, Matthias MS, Bair MJ, Kroenke K. The influence of patient sex, provider sex, and sexist attitudes on pain treatment decisions. *J Pain.* 2014;15(5):551–559.

68   Tori ME, Larochelle MR, Naimi TS. Alcohol or benzodiazepine co-involvement with opioid overdose deaths in the United States, 1999–2017. *JAMA Netw Open.* 2020;3(4):e202361; Opioid overdose deaths by gender. Kaiser Family Foundation. https://www.kff.org/other/state-indicator/opioid-overdose -deaths-by-gender/?currentTimeframe=0&sortModel=%7B%22colId%22:% 22Location%22,%22sort%22:%22asc%22%7D. Accessed June 9, 2021; Sex and gender differences in substance use. NIDA. April 13, 2021.

https://www.drugabuse.gov/publications/research-reports/substance-use-in -women/sex-gender-differences-in-substance-use. Accessed June 9, 2021; Sad- hasivam S, Chidambaran V, Olbrecht VA, et al. Opioid-related adverse effects in children undergoing surgery: unequal burden on younger girls with higher doses of opioids. *Pain Med.* 2015;16(5):985–997; Hernandez-Avila CA, Roun- saville BJ, Kranzler HR. Opioid-, cannabis- and alcohol-dependent women show more rapid progression to substance abuse treatment. *Drug Alcohol Depend.* 2004;74(3):265–272.

69   Goodnough A. Overdose deaths have surged during the pandemic, C.D.C. data shows. *New York Times.* April 14, 2021. https://www.nytimes.com/2021 /04/14/health/overdose-deaths-fentanyl-opiods-coronavirus-pandemic.html.

70   Jonas WB, Crawford C, Colloca L, et al. Are invasive procedures effective for chronic pain? A systematic review. *Pain Med.* 2019;20(7):1281–1293.

CHAPTER 9. ALL IN THE HEAD

1   Palamar JJ, Salomone A, Rutherford C, Keyes KM. Extensive under- reported exposure to ketamine among electronic dance music party attendees. *J Gen Intern Med.* 2021;36(1):235–237; Clark JD. Ketamine for chronic pain: old drug new trick? *Anesthesiology.* 2020;133(1):13–15.

2   Vuckovic S, Srebro D, Vujovic KS, Vucetic C, Prostran M. Cannabi- noids and pain: new insights from old molecules. *Front Pharmacol.* 2018;9:1259.

3   Kondrad E, Reid A. Colorado family physicians' attitudes toward medical marijuana. *J Am Board Fam Med.* 2013;26(1):52–60; Ramaekers JG, Hut- ten N, Mason NL, et al. A low dose of lysergic acid diethylamide decreases pain perception in healthy volunteers. *J Psychopharmacol.* 2021;35(4):398–405.

4   Putnam FW, Helmers K, Horowitz LA, Trickett PK. Hypnotizability and dissociativity in sexually abused girls. *Child Abuse Negl.* 1995;19(5): 645–655.

5   Amiri M, Alavinia M, Singh M, Kumbhare D. Pressure pain threshold in patients with chronic pain: a systematic review and meta-analysis. *Am J Phys Med Rehabil.* 2021;100(7):656–674.

6   Lang EV, Benotsch EG, Fick LJ, et al. Adjunctive non-pharmacological analgesia for invasive medical procedures: a randomised trial. *Lancet.* 2000;355(9214):1486–1490.

7   Madden K, Middleton P, Cyna AM, Matthewson M, Jones L. Hypnosis for

pain management during labour and childbirth. *Cochrane Database Syst Rev*. 2016(5):CD009356; Jensen MP, Patterson DR. Hypnotic approaches for chronic pain management: clinical implications of recent research findings. *Am Psychol*. 2014;69(2):167–177; Thompson T, Terhune DB, Oram C, et al. The effectiveness of hypnosis for pain relief: a systematic review and meta-analysis of 85 controlled experimental trials. *Neurosci Biobehav Rev*. 2019;99:298–310.

8   Bishop GH. The skin as an organ of senses with special reference to the itching sensation. *J Invest Dermatol*. 1948;11(2):143–154; Han L, Dong X. Itch mechanisms and circuits. *Annu Rev Biophys*. 2014;43:331–355.

9   Evers AWM, Peerdeman KJ, van Laarhoven AIM. What is new in the psychology of chronic itch? *Exp Dermatol*. 2019;28(12):1442–1447.

10   Schut C, Grossman S, Gieler U, Kupfer J, Yosipovitch G. Contagious itch: what we know and what we would like to know. *Front Hum Neurosci*. 2015;9:57; Darragh M, Booth RJ, Koschwanez HE, Sollers J, III, Broadbent E. Expectation and the placebo effect in inflammatory skin reactions: a randomised-controlled trial. *J Psychosom Res*. 2013;74(5):439–443; Evers AWM, Peerdeman KJ, van Laarhoven AIM. What is new in the psychology of chronic itch? *Exp Dermatol*. 2019;28(12):1442–1447.

11   Werbart A. "The skin is the cradle of the soul": Didier Anzieu on the skin-ego, boundaries, and boundlessness. *J Am Psychoanal Assoc*. 2019; 67(1):37–58.

12   Werbart A. "The skin is the cradle of the soul": Didier Anzieu on the skin-ego, boundaries, and boundlessness. *J Am Psychoanal Assoc*. 2019; 67(1):37–58.

13   Naldi L, Mercuri SR. Chronic pruritus management: a plea for improvement—can itch clinics be an option? *Dermatology*. 2010;221(3): 216–218.

14   Liebenberg L. Persistence hunting by modern hunter-gatherers. *Curr Anthropol*. 2006;47(6):1017–1026.

15   Tipton CM. The history of "exercise is medicine" in ancient civilizations. *Adv Physiol Educ*. 2014;38(2):109–117; Luque-Suarez A, Martinez-Calderon J, Falla D. Role of kinesiophobia on pain, disability and quality of life in people suffering from chronic musculoskeletal pain: a systematic review. *Brit J Sports Med*. 2019;53(9):554–559.

16   Luque-Suarez A, Martinez-Calderon J, Falla D. Role of kinesiophobia on pain, disability and quality of life in people suffering from chronic musculoskeletal pain: a systematic review. *Brit J Sports Med*. 2019;53(9):554–559.

17   Geneen LJ, Moore RA, Clarke C, Martin D, Colvin LA, Smith BH. Physical activity and exercise for chronic pain in adults: an overview of Cochrane Reviews. *Cochrane Database Syst Rev.* 2017;1:CD011279; Rice D, Nijs J, Kosek E, et al. Exercise-induced hypoalgesia in pain-free and chronic pain populations: state of the art and future directions. *J Pain.* 2019;20(11):1249–1266.

18   Fallon N, Roberts C, Stancak A. Shared and distinct functional networks for empathy and pain processing: a systematic review and meta-analysis of fMRI studies. *Soc Cogn Affect Neurosci.* 2020;15(7):709–723; Rosenthal–von der Pütten AM, Schulte FP, Eimler SC, et al. Investigations on empathy towards humans and robots using fMRI. *Comput Hum Behav.* 2014;33:201–212; Meyer ML, Masten CL, Ma Y, et al. Empathy for the social suffering of friends and strangers recruits distinct patterns of brain activation. *Soc Cogn Affect Neurosci.* 2013;8(4):446–454; Pan Y, Cheng X, Zhang Z, Li X, Hu Y. Cooperation in lovers: an fNIRS-based hyperscanning study. *Hum Brain Mapp.* 2017;38(2):831–841.

19   Bissell DA, Ziadni MS, Sturgeon JA. Perceived injustice in chronic pain: an examination through the lens of predictive processing. *Pain Manag.* 2018;8(2):129–138; Henry SG, Bell RA, Fenton JJ, Kravitz RL. Goals of chronic pain management: do patients and primary care physicians agree and does it matter? *Clin J Pain.* 2017;33(11):955–961.

20   Goldstein P, Weissman-Fogel I, Dumas G, Shamay-Tsoory SG. Brain-to-brain coupling during handholding is associated with pain reduction. *Proc Natl Acad Sci USA.* 2018;115(11):E2528–E2537.

21   Wolters F, Peerdeman KJ, Evers AWM. Placebo and nocebo effects across symptoms: from pain to fatigue, dyspnea, nausea, and itch. *Front Psychiatry.* 2019;10:470; Kirchhof J, Petrakova L, Brinkhoff A, et al. Learned immunosuppressive placebo responses in renal transplant patients. *Proc Natl Acad Sci USA.* 2018;115(16):4223–4227.

22   Rief W, Avorn J, Barsky AJ. Medication-attributed adverse effects in placebo groups: implications for assessment of adverse effects. *Arch Intern Med.* 2006;166(2):155–160; Bingel U, Wanigasekera V, Wiech K, et al. The effect of treatment expectation on drug efficacy: imaging the analgesic benefit of the opioid remifentanil. *Sci Transl Med.* 2011;3(70):70ra14; Kam-Hansen S, Jakubowski M, Kelley JM, et al. Altered placebo and drug labeling changes the outcome of episodic migraine attacks. *Sci Transl Med.* 2014;6(218):218ra215; Holtedahl R, Brox JI, Tjomsland O. Placebo effects in trials evaluating 12 selected

minimally invasive interventions: a systematic review and meta-analysis. *BMJ Open*. 2015;5(1):e007331.

23 Kaptchuk TJ, Kelley JM, Conboy LA, et al. Components of placebo effect: randomised controlled trial in patients with irritable bowel syndrome. *BMJ*. 2008;336(7651):999–1003.

24 Use of placebo in clinical practice. Code of medical ethics opinion 2.1.4. American Medical Association. https://www.ama-assn.org/delivering-care /ethics/use-placebo-clinical-practice. Accessed June 11, 2021; Linde K, Atmann O, Meissner K, et al. How often do general practitioners use placebos and non-specific interventions? Systematic review and meta-analysis of surveys. *PLoS One*. 2018;13(8):e0202211.

25 Carvalho C, Caetano JM, Cunha L, Rebouta P, Kaptchuk TJ, Kirsch I. Open-label placebo treatment in chronic low back pain: a randomized controlled trial. *Pain*. 2016;157(12):2766–2772.

26 von Wernsdorff M, Loef M, Tuschen-Caffier B, Schmidt S. Effects of open-label placebos in clinical trials: a systematic review and meta-analysis. *Sci Rep*. 2021;11(1):3855.

27 Canovas L, Carrascosa AJ, Garcia M, et al. Impact of empathy in the patient-doctor relationship on chronic pain relief and quality of life: a prospective study in Spanish pain clinics. *Pain Med*. 2018;19(7):1304–1314.

28 Bylund CL, Makoul G. Empathic communication and gender in the physician-patient encounter. *Patient Educ Couns*. 2002;48(3):207–216.

29 Hojat M, Gonnella JS, Nasca TJ, Mangione S, Vergare M, Magee M. Physician empathy: definition, components, measurement, and relationship to gender and specialty. *Am J Psychiat*. 2002;159(9):1563–1569.

30 Han S. Neurocognitive basis of racial ingroup bias in empathy. *Trends Cogn Sci*. 2018;22(5):400–421.

31 Cao Y, Contreras-Huerta LS, McFadyen J, Cunnington R. Racial bias in neural response to others' pain is reduced with other-race contact. *Cortex*. 2015;70:68–78.

32 Sheng F, Han S. Manipulations of cognitive strategies and intergroup relationships reduce the racial bias in empathic neural responses. *Neuroimage*. 2012;61(4):786–797.

33 Bonica JJ. Organization and function of a multidisciplinary pain clinic. In: Weisenberg M, Tursky B, eds. *Pain: New Perspectives in Therapy and Research*. Boston: Springer US; 1976:11–20.

34 Kamper SJ, Apeldoorn AT, Chiarotto A, et al. Multidisciplinary biopsycho-

social rehabilitation for chronic low back pain: Cochrane systematic review and meta-analysis. *BMJ*. 2015;350:h444; Kligler B, Bair MJ, Banerjea R, et al. Clinical policy recommendations from the VHA State-of-the-Art Conference on non-pharmacological approaches to chronic musculoskeletal pain. *J Gen Intern Med*. 2018;33(Suppl 1):16–23.

35 Schatman ME, Webster LR. The health insurance industry: perpetuating the opioid crisis through policies of cost-containment and profitability. *J Pain Res*. 2015;8:153–158.

36 Schatman ME, Webster LR. The health insurance industry: perpetuating the opioid crisis through policies of cost-containment and profitability. *J Pain Res*. 2015;8:153–158.

37 Lebovits A. Maintaining professionalism in today's business environment: ethical challenges for the pain medicine specialist. *Pain Med*. 2012;13(9): 1152–1161.

38 VA Office of Public and Intergovernmental Affairs. VA becomes first hospital system to release opioid prescribing rates. US Department of Veterans Affairs. January 11, 2018. https://www.va.gov/opa/pressrel/pressrelease .cfm?id=3997. Accessed June 11, 2021; Mattocks K, Rosen MI, Sellinger J, et al. Pain care in the Department of Veterans Affairs: understanding how a cultural shift in pain care impacts provider decisions and collaboration. *Pain Med*. 2020;21(5):970–977.

39 Taheri AA, Foroughi AA, Mohammadian Y, et al. The effectiveness of acceptance and commitment therapy on pain acceptance and pain perception in patients with painful diabetic neuropathy: a randomized controlled trial. *Diabetes Ther*. 2020;11(8):1695–1708.

40 Coronado RA, Brintz CE, McKernan LC, et al. Psychologically informed physical therapy for musculoskeletal pain: current approaches, implications, and future directions from recent randomized trials. *Pain Rep*. 2020;5(5):e847; Trindade IA, Guiomar R, Carvalho SA, et al. Efficacy of online-based acceptance and commitment therapy for chronic pain: a systematic review and meta-analysis. *J Pain*. April 20, 2021. https://doi.org /10.1016/j.jpain.2021.04.003.

41 Godfrey E, Wileman V, Galea Holmes M, et al. Physical therapy informed by acceptance and commitment therapy (PACT) versus usual care physical therapy for adults with chronic low back pain: a randomized controlled trial. *J Pain*. 2020;21(1–2):71–81.

42 Hamilton A. Stop telling chronic pain patients that we should just accept our pain. Rooted in Rights. June 21, 2018. https://rootedinrights.org/stop-telling

-chronic-pain-patients-that-we-should-just-accept-our-pain. Accessed June 11, 2021.

43 Godfrey E, Wileman V, Galea Holmes M, et al. Physical therapy informed by acceptance and commitment therapy (PACT) versus usual care physical therapy for adults with chronic low back pain: a randomized controlled trial. *J Pain*. 2020;21(1–2):71–81.

44 Munkholm K, Paludan-Muller AS, Boesen K. Considering the methodological limitations in the evidence base of antidepressants for depression: a reanalysis of a network meta-analysis. *BMJ Open*. 2019;9(6):e024886.

45 Bell JT, Loomis AK, Butcher LM, et al. Differential methylation of the TRPA1 promoter in pain sensitivity. *Nat Commun*. 2014;5:2978.

INDEX

293

heat
TRP (receptors), 49
as type of harm, 48–49
helplessness in pain catastrophizing, 89
Henig, Robin Marantz, 166
hepatitis, C, 174
Heraclides Ponticus, 134
heroin, 8, 142, 156, 157, 159, 161, 171,
174, 186, 188, 202
hippocampus
adult hippocampal neurogenesis, 126
PKMzeta and, 122–123
roles, 122, 124
Hippocrates
disciples/opium and, 134–135
as physician/oath of, 108, 134, 135
History of Corporal Punishment, The
(Scott), 193
HIV/AIDS, 47, 149–150, 174
homeostasis (body's physiologic
processes), 68, 190
Hong Kong, 137
hope significance, 106
hospice creation, 8, 157–159, 160, 169
Hugo, Victor, 98
Huxley, Aldous, 164
hyperalgesia, 169
hyperkatifeia, 179
hypnosis/hypnotherapy
effects, 229, 231, 232
goal, 231
self-hypnosis, 231–232
hypodermic syringe
invention, 140
racism/xenophobia and, 141
as status symbol, 141
use, 139, 140–141, 154–155
hypothalamus, 240

iatrogenesis, 145, 185
"idiopathic" meaning, 106
Illich, Ivan
background, 144–145
*Medical Nemesis: The Expropriation of
Health*, 145, 171
views/mass medicine and drugs, 145–
146, 147, 171, 245
incarceration and inequalities, 201–202
Index Medicus, 113

Industrial Revolution, 139, 142–143
insula
anterior insula/role, 62, 120
consciousness and, 59
emotions and, 63
pain and, 59, 61
posterior insula/role, 59, 61, 62
role, 59, 62–63
self-awareness and, 62–63
sensory-motor pathway and, 59
von Economo neurons, 63
insular cortex
limbic system and, 53
pain/pain center and, 61–62
See also insula
insurance, 47, 79, 85, 94, 192, 227,
246
integrated information theory, 20–21
*International Association for the Study of
Pain* (IASP), 73
invertebrates
nociception and, 25
pain and, 26–27
Isnard, Jean
brain pain center and, 60–62
epilepsy/patient Jacob, 60–61
itch/scratching cycle, 233–234, 235

Jahangir, Mughal emperor, 136
James VI, King, 208
Jamison, Bob
as pain psychologist/work, 88–89,
91–92
position/office, 88–89
Jaynes, Julian, 33
Jesus, 194
*Journal of the American Medical
Association*, 164, 197
Julius, David, 49
Jung, Carl, 1

Kahneman, Daniel, 125
Kaptchuk, Ted
background, 241
placebo response/physician effects on
patients and, 241–243
Kefauver, Estes
Mafia prosecution and, 167
resegregation views, 167

Rabail Baig

Haider Warraich is a doctor at Brigham and Women's Hospital and in the VA Boston Healthcare System and an assistant professor at Harvard Medical School. He is the author of *Modern Death* and *State of the Heart* and regularly writes for the *New York Times* and *Washington Post*, among others. He lives in Wellesley, Massachusetts.